A PRACTICAL GUIDE TO INSPECTING

PLUMBING

By Roy Newcomer

CONTENTS

INTRODUCTION i

Chapter One

THE PLUMBING INSPECTION 1

    Inspection Guidelines and Overview 1

Chapter Two

WATER SUPPLY ENTRANCE .5

    Public Water Supply 5

    Private Water Supply 10

    Inspecting Private Water Supply 15

    Reporting Your Findings .19

    Filling in Your Report 20

    Worksheet 22

Chapter Three

WATER SUPPLY PIPING 23

    Corrosion 23

    Types of Supply Piping 25

    Inspecting Supply Piping .29

    Functional Flow 34

    Cross Connections 36

    Reporting Your Findings .37

    Worksheet 39

Chapter Four

DRAIN, WASTE, AND VENT SYSTEM 40

    Waste Disposal 41

    The Venting System 43

    DWV Piping Materials 45

    Inspecting DWV Piping 48

    Inspecting Drainage 52

    Reporting Your Findings .55

    Worksheet 58

Chapter Five

HOT WATER SYSTEM 59

    Water Heaters 59

    Gas Water Heaters 62

    Electric Water Heaters 68

    Other Hot Water Systems 69

    Reporting Your Findings .70

    Worksheet 71

Chapter Six

FIXTURES AND FAUCETS 72

    Inspecting Faucets 72

    Inspecting Sinks, Tubs, and Showers 74

    Inspecting Toilets 77

    Putting It All Together 79

    Reporting Your Findings .80

Chapter Seven

    Gas Piping 81

EXAM 83

GLOSSARY 89

INDEX 93

INTRODUCTION

My background includes many years in construction and several more as the owner of a Century 21 real estate franchise. In 1989, I started a home inspection company that grew larger than I ever imagined it could. Training my own staff of inspectors to the highest inspection standards led to my teaching home inspection seminars across the country and developing study courses, books, and videos for home inspectors. The American Home Inspectors Training Institute was founded as a result of my desire to share this experience and knowledge in home inspection.

The *Practical Guide to Inspecting* series is intended for both beginning and experienced home inspectors. So if you're studying home inspection for the first time or are using the materials as a refresher, these guides should be of assistance to you.

I've created these guides to include all aspects of home inspection. Not only a broad technical background in home systems, but the other things you need to know in order to perform a *good* inspection of those systems. They lay out technical information, guidelines for the inspection, how-to instructions for inspecting system components, and the defects, deficiencies, and problems you'll be looking for during the inspection. I've also included some advice on how to report your findings to the home inspection customer.

I've been a member of several professional organizations for a number of years, including ASHI® (American Society of Home Inspectors), NAHI™ (National Association of Home Inspectors), and CREIA® (California Real Estate Inspection Association). I am a great supporter of those organizations' quest to promote excellence in home inspection.

I encourage you to follow the standards of the organization to which you might belong, or any state regulation that might take precedent over the standards used here. Use the standards in this book as a general guide for study and apply the standard or state regulation that applies to you.

The inspection guidelines presented in the Practical Guides are an attempt to meet or exceed standards and regulations as they exist at the revision date of the guides.

There's a lot to learn about home inspection. For beginning inspectors, there are some *hands-on exercises* in this guide that should be done. I'm a great believer in learning by doing, and I hope you'll try them. There are also some of my *personal inspection stories* to let you know what it's really like out there.

The *inspection photos* referenced in this text can also be found on www.ahit.com/photos. You'll read the story about each one as you go along. Be sure to watch for my *Don't Ever Miss* lists. I've included them to alert home inspectors to report those defects (if found during the inspection) in the inspection report. If missed, these items are often the cause for lawsuits later. Finally, to help you see how you're doing as you study this guide, I've included some *worksheets*. The answers are given for each one for self checking. Give them a try. Checking yourself can help you lock important information in your mind. There's also a *final exam* that you can complete and send in to us. Many organizations and states have approved this book for continuing education credits. Submit the exam with the required fee if you need these credits.

In total, the *Practial Guide to Inspecting* series covers all aspects of the general home inspection. Each guide covers a major aspect of the inspection, as their titles show:

Electrical
Exteriors
Heating and Cooling
Interiors, Insulation, Ventilation
Plumbing
Roofs
Structure

If you are interested in other titles in the series, please call us at the American Home Inspectors Training Institute to order them. Call toll free at 1-800-441-9411.

Roy Newcomer

INSPECTING PLUMBING

Chapter One

THE PLUMBING INSPECTION

As with the electrical system, much of the plumbing system is hidden behind the walls of the home. The home inspector performs a **visual inspection** of the exposed supply and waste distribution piping and fixtures, reporting any defects found in the system and helping customers to understand those defects.

Inspection Guidelines and Overview

These are the standards of practice that apply to the plumbing system. The chart continues on the next page.

Plumbing System	
OBJECTIVE	To identify major deficiencies in the interior plumbing system.

OBSERVATION	<u>Required to inspect and report:</u>
	• Interior water supply and distribution system including:
	  — Piping materials, supports, insulation
	  — Fixtures and faucets
	  — Functional flow and leaks
	  — Cross connections
	• Interior drain, waste, and vent system:
	  — Traps
	  — Drain, waste, and vent piping
	  — Piping supports and pipe insulation
	  — Leaks and functional drainage
	• Hot water systems:
	  — Water heating equipment
	  — Normal operating controls
	  — Automatic safety controls
	  — Chimney, flues, and vents
	• Fuel storage and distribution systems:
	  — Interior fuel storage equipment, supply piping, venting, and supports
	  — Leaks
	• Sump pumps
	• Seismic bracing (where applicable)
	<u>Not required to observe:</u>
	• Water conditioning systems
	• Fire and lawn sprinkler systems
	• Onsite water supply quantity and quality
	• Spas, except as to functional flow and drainage

Guide Note

Pages 1 to 4 outline the content and scope of the plumbing inspection. It's an overview of the inspection, including what to observe, what to describe, and what specific actions to take during the inspection. Study these guidelines well.

ACTION	Required to:
	• Operate all plumbing fixtures.
	Not required to:
	• State the effectiveness of anti-siphon devices.
	• Determine whether water supply and waste disposal systems are public or private.
	• Operate automatic safety controls.
	• Operate any valve except water closet flush valves, fixture faucets, and hose faucets.

This table provides a good outline of the guidelines to govern the plumbing inspection. Not every single detail is presented in the table, but enough of an overview to have a good idea of what is inspected and what is not required to be inspected. Here is a layout of what the plumbing inspection entails:

- **Water supply entrance:** Although most standards of practice state that the home inspector is not required to determine if the water supply is public or private, we suggest that you do. The home inspector should identify the service piping and inform the customer if this piping is lead. For a public service, the inspector should locate the water meter and main shut-off valve. The inspector doesn't test the main shut-off valve. In fact, the home inspector is **not required to operate any valve except toilet flush valves and faucets**.

For a private service, the home inspector locates the source of water supply and determines the type of well and the condition of its well equipment. The inspector checks the pressure gauge and watches for a waterlogged tank.

Note that the home inspector is **not required to inspect water conditioning systems** and is not required to test or report on the **quantity or quality** of the water supply.

Supply distribution piping: The home inspector identifies the materials used in distribution piping. All visible interior hot and cold water piping is inspected for leaking and deterioration. The home inspector looks for evidence of leaking throughout the house and keeps an eye out for any repairs to the distribution system. The home inspector watches for and reports all safety hazards such as cross connections and the presence of asbestos insulation on piping.

The home inspector is **not required to inspect fire and lawn sprinkler systems** or **spas**, except to functional flow and drainage. Also, the inspector is not required to determine the **effectiveness of anti-siphon devices**.

- **Fixtures and faucets:** All fixtures and faucets in the supply system are operated, paying attention to the

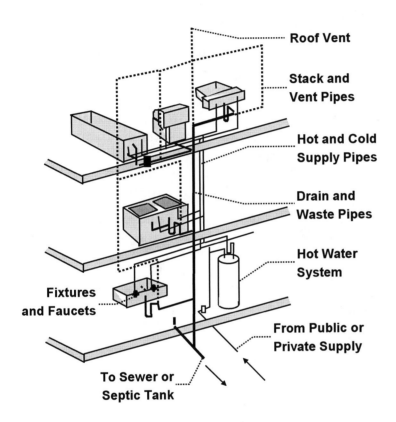

Roof Vent

Stack and Vent Pipes

Hot and Cold Supply Pipes

Drain and Waste Pipes

Hot Water System

Fixtures and Faucets

From Public or Private Supply

To Sewer or Septic Tank

functional flow of water and leaking. This includes sinks, toilets, tubs and showers, and utility room tubs.

- **Drain, waste, and vent system:** The home inspector identifies the piping material in this system and inspects it for leaks and deterioration. The inspector determines if the proper traps are used in the system and checks the visible piping for support and insulation. Drains are checked for functional drainage. The home inspector verifies the

NOT REQUIRED TO

- State the effectiveness of anti-siphon devices
- Operate automatic safety controls
- Operate any valve except toilet flush valves and faucets
- Observe water conditioning systems
- Observe fire and lawn sprinkler systems
- Observe water supply quantity or quality
- Observe spas, except for flow and drainage

existence of the vent system and its emergence through the roof. Any indication of leaking sewer gas is reported.

The inspection of a private waste disposal system is beyond the scope of the general home inspection. The home inspector is **not required to inspect the septic tank** or to report on it other than to mention its existence and to advise customers to have the tank pumped out on a regular basis.

- **Hot water system:** The home inspector always inspects the water heater during the plumbing inspection. The inspector describes the **type** of water heater as gas, electric, oil, or other types, noting its **brand name, capacity, and approximate age.** Customers are informed of the remaining useful lifetime of the water heater. The home inspector checks for the presence and proper installation of a safety relief valve and extension. Note again that the inspector is **not required to operate any valves or operate automatic safety controls.** The tank itself is inspected for condition. The water heater (gas only) is turned up, and the home inspector listens to its operation. The burner cover is removed to observe the flame in gas water heaters. The home inspector checks for proper venting.

- **Fuel storage and distribution:** The inspection of interior gas piping within the plumbing inspection is included here. That's because it's convenient to check gas piping while inspecting the water piping. The inspection includes the location of the meter or bottled gas tank. The most important part of this inspection is noticing the odor of a gas leak, which must be reported as a safety hazard.

- **Sump pumps:** In this guide, we're going to be studying "gray water" **sanitary pumps** that may be used to pump water from clothes washers into the waste disposal system. The home inspector will inspect such sanitary pumps. (The inspection of sump pumps is also presented in *A Practical Guide to Inspecting Structure*, referring to fresh water sumps that pump water from the foundation drainage system.)

Chapter Two

WATER SUPPLY ENTRANCE

The first component of the plumbing system to be studied is the water supply entrance, although it isn't necessarily the first to be inspected during a real home inspection. We'll look at the components one by one before talking about inspection order.

The inspection of the water supply begins at the pipe that brings water into the house. This pipe is called the **service pipe** or the **house main**, and the inspector describes the material it is made of. The inspector also checks for the presence of the **main shut-off valve** on the service pipe. The inspector should determine whether the water supply comes from a **public service or private source** such as a well on the property. The water supply entrance inspection includes the following:

- **Public or private** water source
- The **service pipe** material
- **Meter location**, if public supply
- The presence of **main shut-off valve** and its location
- The **pump and well tank** condition and the operation of the **pressure gauge**, if private supply

Public Water Supply

A service line carries water from the city water main at the street to the house. The **minimum service pipe size** for bringing water into a house from a public water service is typically 3/4" ID (inside diameter), but the home inspector may see 1" and larger pipes in old installations, especially in old inner city areas and in new construction of large homes. The service pipe is usually buried far below grade to prevent freezing in cold climates or interference with surface plantings in warmer climates. It may surface into the house through the foundation wall or the basement floor. City water is usually provided at a **pressure** of 40 to 70 psi (pounds per square inch). Higher pressures will put stress on the household water system, while lower pressures will not provide enough pressure to the house.

The public service line may be made of one of the following materials:

INSPECTING WATER SUPPLY ENTRANCE

- Public or private water source

- Service pipe material

- Meter location

- Location of main shut-off valve

- Condition of private pump and well tank and operation of pressure gauge

Guide Note

Pages 5 to 22 present the study and inspection of the water supply entrance portion of the plumbing system.

The house main or service pipe is the pipe that brings water from its public or private source into the house.

The main shut-off valve turns off the water supply to the house. The use of the abbreviation ID with pipe measurements refers to the inside diameter of the pipe.

The abbreviation psi is a measurement of water pressure and means pounds per square inch.

- **Copper:** This is the most commonly used material for service pipes today. Since 1970, pipes are usually 3/4" ID, although 1/2" ID was also used in the 20 years before 1970. Type K, or thick-walled copper pipes, are used for such underground installations, and last indefinitely.

- **Galvanized steel:** Galvanized steel is steel coated with a corrosion-resistant layer of zinc. Galvanized steel is not used very often as service lines, and those communities that used it may require pipe of at least 1 1/4" ID. These pipes can last 30 to 60 years, depending on the acidity of the water, before the zinc is lost and the steel begins to rust.

- **Lead:** Lead piping was used between the street main and the house up until the 1950's, and many of these lines are still in use. There is concern over the safety of these lead service lines because of the lead which can contaminate the water.

- **Plastic:** Public water supply lines are usually not plastic, although plastic piping is often used in private water systems. If **polyethylene (PE)** is used in high-pressure city systems, it may burst after only a few years. **Polybutylene (PB)**, which is a blue color, has also been used. But even PB shrinks in cold weather and can pull loose from fittings, can split from temperature changes, or can become brittle from the chlorination of the water.

With public water services a **water meter** may be present either inside or outside the house. In some communities, there are un-metered services and no meter may be found, although this is not common. When street pressure is too high for home use, there may be a **pressure reducing valve** (PRV) on the service pipe near the meter.

NOTE: In rare cases, where water pressure from city service is too low, a **booster pump and pressure tank** may be installed. This setup looks like a shallow well system, but the presence of a meter should confirm that the water supply is public.

Each house should have a **main shut-off valve** on the house side of the meter to turn off the water supply to the house. This valve should be operable by hand not by wrench, so anyone can turn it off. Homeowners should tag the main shut-off and let everyone in the house know where it is located so it can be turned off in case of a catastrophic plumbing leak This valve may also be called the main house valve or the main valve.

When inspecting the water supply entrance, first locate the incoming service pipe. Identify the service as public or private based on the presence of a water meter and lack of well equipment. During the inspection, watch for the following:

- **Lead service pipes:** Both lead and galvanized steel pipes are gray in color. To determine if the service line is lead, use your screwdriver to make a scratch in the pipe. Lead is soft and the scratch can be made easily. The scratch line will be shiny. Galvanized steel is harder to scratch, and the resulting scratch line is not as shiny as lead. Lead pipes can be easily bent. Galvanized pipe is much harder to bend and usually runs in straight lengths. Some home inspectors carry a magnet to help with this identification. Galvanized steel is magnetic, but lead isn't. Another difference is that galvanized steel pipes are threaded, while lead pipes have a distinctive tulip bulb shaped joint.

The home inspector should let the customer know when lead pipes have been identified and it should be noted in the inspection report. Because of health concerns, customers may want to have the water tested for lead contamination. The inspector can suggest that water be run for 2 or 3 minutes before using it for drinking or that the customers use bottled water for consumption.

LEAD vs. GALVANIZED STEEL		
	Lead	Steel
• Joints	Bulb	Threaded
• Magnetic	No	Yes
• Scratch	Easy, shiny	Harder, dull
• Bends	Easy	Harder

#1 Lead service pipe

*Photo #1 shows a **lead service pipe** coming out of the basement floor to the house (see at right in photo). This pipe can be identified as lead by the characteristic bulging or tulip bulb shaped joint, which you can see between the entry and the meter. Here, this piping is lead up to the first horizontal portion before the meter. We let the customer know that this is a lead service pipe and noted it in the report. We also suggested they may want a water quality test to see how much lead is in the water. Don't ever miss identifying a lead service pipe!*

*Photo #2 shows **another lead service pipe** coming out of the basement floor into the house. Notice the pipe's dull gray color and how it curves as it reaches over toward the main stack. Lead pipe is soft enough to bend in this way, while galvanized steel isn't. A scratch test confirmed that this pipe is lead. Again, we let the customer know that the water supply lines coming in were lead. Notice the lever just to the left of the lead pipe. This is the main shut-off for the house. It's a lever connected to the pipe below floor level. It's pulled to the side to turn off the water.*

#2 Another lead service pipe

INSPECTING PUBLIC SUPPLY

- Lead service pipe
- Pipe too small
- Missing, damaged, or leaking main shut-off valve

For Beginning Inspectors

Locate the service pipe into your own home. Perform the inspection as described in these pages. First, determine whether the source of water is public or private. Note the pipe material and use the scratch test if necessary to identify lead piping. Locate the main shut-off valve. You might want to tag this valve and show your family where it is in case of an emergency.

- **Pipe too small:** Check out the size of the service pipe. Copper pipe should be 3/4", and if you find 1/2" pipe, it should be pointed out to the customer as it can cause water flow problems. Plastic pipe may be used in your area. This should also be at least 3/4".

- **Missing, damaged, or leaking main shut-off valve:** Check for a main shut-off valve *on the house side of the meter*, and be sure to report if the valve is missing. Even if the water meter is outside, the main shut-off may be inside the house. Inspect the valve for damage or leaking. It should be operational by hand in case of an emergency. Take another look at **photo #2** as a reminder that sometimes the main shut-off may not be the traditional sort shown above.

Do not test the main shut-off valve by turning it off. Too often, a home inspector turns off a main shut-off valve only to find that it can't be turned back on or a valve that hasn't been operated for a long time becomes damaged or begins leaking after the test. If you damage it, you'll pay to have a plumber come in to fix it. Your job is to let the customer know where the valve is located, what its function is, and how it *would be* operated. If the valve is already damaged or leaking, suggest that it be repaired.

There may be another shut-off visible on the street side of the meter which allows water to be turned off by the water service in order to work on the meter. This valve may have to be operated with a wrench. It's not intended to serve as the homeowner's shut-off. See **Photo #1** for an example.

#3 Metered water service

*Photo #3 shows a **metered water service**. Here, there is a copper service pipe entering the house. Although it can't be easily seen in this photo, there is a turn-off before the meter that must be operated with a wrench. But notice that there isn't any hand-operated main shut-off on the house side of the meter. We reported the missing main shut-off.*

NOTE: It's when you're inspecting the water supply entrance in the basement that you'll be looking for the electrical grounding conductor (see *A Practical Guide to Inspecting Electrical*). The grounding conductor should be connected to the street side of the water meter and there should be a jumper across the meter for effective grounding. Of course, a plastic service line cannot be used for electrical grounding.

WELL COMPONENTS
- The well
- Well pump
- Pressure tank
- Pressure switch and pressure gauge
- Service pipe and main shut-off valve

Guide Note

Inspecting the home for proper grounding is presented in another of our guides — A Practical Guide to Inspecting Electrical.

Definitions

An aquifer is a water-bearing strata of permeable rock, sand, or gravel.

An artesian well is one whose aquifer has enough internal pressure to bring water to the surface without a pump.

The use of the abbreviation gpm is a measurement of water flow and means gallons per minute.

Private Water Supply

A private water supply system typically provides water to the home from a **well**, although water may also be obtained from a pond, stream, or spring. The components of a well system are as follows:

- The **well** itself
- The **well pump** that brings the water to the surface
- The **service pipe** between the well and the pressure tank
- A **main shut-off valve** between the well and the pressure tank
- A **pressure or storage tank** that holds a supply of water
- A **pressure switch** that activates the pump when water pressure in the tank falls and deactivates the pump when pressure resumes an acceptable level
- A **pressure gauge** that gives a readout of the water pressure within the tank

The home inspector can generally identify a water supply as private by the presence of the pressure tank in the home (or pump house) *and* the absence of a water meter. (See note on page 6 about the use of a booster pump and pressure tank in rare cases with a public water service.)

A private water supply is usually provided from an underground stream or **aquifer** as opposed to surface water. Surface water seeps into the ground until it reaches an impermeable layer of rock, then it flows underground along this rock strata until it reaches a river or the ocean. This layer of permeable water-containing rock is the aquifer. Wells are dug or driven to reach the aquifer, which in some areas may be less than 25' below grade, up to several hundred feet below grade. For example, drilled wells in Connecticut are required to be at least 450' deep, while depths of 1100' are not uncommon. An **artesian well** is one whose aquifer is under enough natural pressure to force water from the ground without a pump.

Wells up to 25' deep are called **shallow wells**. Wells more than 25' deep are called **deep wells**. Although rules regarding the location of a private well on a property vary from community to community, the well is usually required to be at least 75' from a private waste disposal system or septic tank (some areas require up to 200'), should be located uphill of a contaminating source, and a required distance from a stream or pond. Typically, the first 50' of the well must be protected by an impervious casing

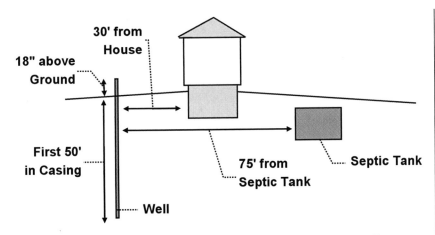

sealed at its lower end. Its upper end should be 12" to 18" above the ground to prevent ground water from entering. A distance of 30' from the house is considered safe if chlordane is or was used to control termite activity.

The purpose of a **well pump** is to draw water from the well and push it through the house piping system with enough force to provide an adequate flow. A flow rate of 5 to 7 gpm (gallons per minute) is considered adequate in most areas of the country. A peak flow rate of 10 gpm is the optimum flow rate for a modern home with 2 bathrooms. The following types of well pumps are used:

- **Piston pump:** This type of pump is used in older systems and is no longer installed today. It is a **reciprocating pump**, which is a motorized version of the old hand pump. A motor drives a series of pistons that lift water higher and higher, discharging it on every other stroke. The piston pump may be used with shallow wells, where the motor and piston assembly will both be above the ground. When used with deep wells, the motor is above the ground and the piston assembly is in the well.

- **Jet pump:** The jet pump consists of a **centrifugal pump** and a jet assembly. The centrifugal pump can be thought of as a small paddle wheel driven by a motor. The pump recirculates pressurized water into the well, creating a suction that draws more well water into the jet assembly and pushes this water up to the pump. Some of the water enters the house distribution piping, while the rest is diverted back to the jet assembly.

When the jet pump is used with a **shallow well,** the home inspector will see **only 1 pipe** extending to the well. In

Personal Note

"One of the instructors here at the Institute was conducting an inspection and was checking the distance between the well riser in the front yard and the septic tank in the back. He noticed that the neighbor had reversed the locations with his well in the back yard and the septic in the front. The well under inspection was the correct distance from its own septic system, but was only 20' from the neighbor's.
"Either the homeowner or the neighbor would have to switch the systems so distances were correct in all directions."

Roy Newcomer

this case, the jet assembly is above ground with the centrifugal pump. When the jet pump is used with a **deep well**, there will be 2 pipes extending to the well. That's how you can tell if the jet pump is used with shallow or deep wells.

In order to operate properly, the deep well jet pump needs to be filled with water and have about 15 to 20 pounds of back pressure. When the water in the casing falls below the level of the jet (when the well is pumped dry), it must be re-primed and pressurized before it operates again.

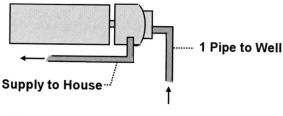

JET PUMP FOR SHALLOW WELL

1 Pipe to Well

Supply to House

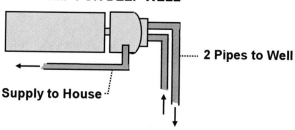

JET PUMP FOR DEEP WELL

2 Pipes to Well

Supply to House

- **Submersible pump:** The submersible pump consists of a motor and an electrically driven **centrifugal pump**, both of which are designed to operate under water. With the submersible pump, both motor and the pump are within the well. Although the submersible pump can be used with shallow wells, it's often used with deep wells.

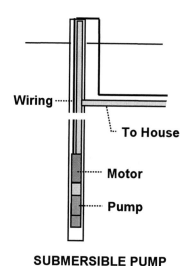

Wiring

To House

Motor

Pump

SUBMERSIBLE PUMP

Water is drawn into the unit through screened openings between the pump and the motor. The submersible pump delivers water to a storage tank through a single pipe and needs no priming.

For Beginning Inspectors

It would be well worth your time to locate a friend with a private well. Take a look at the visible equipment and learn what you can about the type of well pump, the depth of the well, and whatever else you can. Study the motor if above ground, take a look at the pressure tank, and locate the pressure gauge. The more experience you have with the various types of private water systems, the better your inspections will be.

The following chart summarizes the information about the piston, jet, and submersible pumps. Take the time to sort out this information. It's important to know.

Pump	Type	Location	Use
Piston	Reciprocal	With shallow wells, motor and piston above well. With deep wells, motor above and piston in well.	In older homes for both shallow and deep wells, but generally no longer in use.
Jet	Centrifugal	With shallow wells, pump and jet assembly above well. With deep wells, pump above and jet assembly in well.	With both shallow and deep wells. Only 1 pipe to well indicates shallow well; 2 pipes to well indicate deep well.
Submersible	Centrifugal	Both motor and pump located in well.	With both shallow and deep wells, but most often deep.

The **life expectancy** of a well pump can be only 7 to 10 years, although many pumps may last from 20 to 30 years. Submersible pumps generally last about 20 years. In many cases, the electrical box for the pump will have a date code on it that may indicate when the pump was installed.

In the days when pumps were noisy and pressure tanks were large, well equipment was often placed outside the house in separate **pump house** or in an underground **well pit**. The well pit is a masonry chamber with a hatch cover and a floor below the frost line. The home inspector should inspect the well equipment, even if located in the pump house or a well pit, if accessible.

The other main components of a private well system are as follows:

- **Pressure tank:** This storage tank holds water from the well and is usually located on the lower level of the house or basement. Other methods of storing water for home use include an inside reservoir called a **vault**, often located in the attic, or an outside elevated reservoir called a

A well pit is an under-ground chamber, usually made of masonry, that houses the well equipment.

A pump house is a separate building, built for the purpose of housing well equipment.

A vault is an elevated indoor water reservoir, often located in the attic, from which water flows by gravity. A standpipe is an outdoor elevated water reservoir.

standpipe. Water, once in a vault or standpipe, will flow into the house by gravity. There also used to be underground vaults or cisterns, that gathered and held rain water before mechanical pumps came into use.

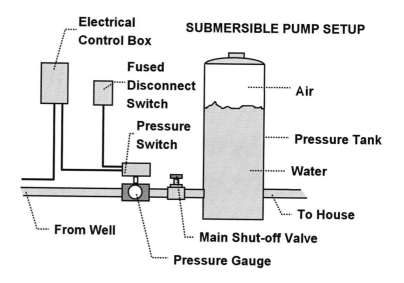

These days, the home inspector is most likely to find water stored in a pressure tank with a capacity of only a few gallons up to 40 or more gallons. The water in the tank is under pressure, and the tank must be partially filled with **air** to maintain this pressure. Normally, the tank is maintained between 1/2 and 2/3 full of water. If the tank becomes filled with water, due to the absorption of the air into the water over time, then the tank becomes **waterlogged**. This causes the pump to cycle rapidly and repeatedly. Some small tanks have a diaphragm separating the air and the water, preventing air from being absorbed. However, there can be problems with the diaphragm which can also cause a waterlogged condition.

The pressure tank should be wrapped or insulated to prevent any corrosion caused by condensation on the sides of the tank. Few homeowners do this.

- **Pressure switch and gauge:** A private well system must have a pressure switch that automatically starts and stops the pump at predetermined pressures. Typical low and high limit settings for water in the pressure tank may be 20 and 40 psi, 30 and 50 psi, or even 40 and 60 psi. If, for example, the low and high are set at 30 psi and 50 psi, then the pressure switch would turn on the pump when pressure levels fall below 30 psi and turn off the pump when levels reach 50 psi. Pressures above 60 psi can harm the supply piping. When the pressure tank becomes waterlogged, the pressure switch can be heard clicking continuously, trying to bring water pressure into the acceptable range.

For a jet pump, the pressure switch is located in the pump assembly. For the submersible pump the switch is above ground near the pressure tank. A **pressure gauge**, which shows the pressure reading of the tank, may be located near or on the tank itself.

#4 Pressure tank

Photo #4 shows a ***pressure tank*** *for a submersible pump. Notice where the service pipe from the well and the well wiring enter the house through the foundation wall? With this pressure tank, you can see evidence of where the water line is on the tank — about the top third of the tank contains air. The rest of the tank shows evidence of condensation. Here, the pressure switch is at the top and just to the right of the tank, which you can tell because of the wiring from the electrical box. The pressure gauge is on the tank. This tank shows water staining on the sides, which is from condensation, not leaking.*

Inspecting Private Water Supply

When inspecting the private water supply entrance, the home inspector should identify the **source** of the water, if possible. When there is well equipment, it should be inspected whether located in the basement, in a pump house, or in a well pit. The inspector should report the **location** of the well equipment in the report. The **type of well pump** should be identified and the well itself as shallow or deep. The home inspector should determine the **age of the well pump** and report whether it may have to be replaced within the next 5 years.

- Turn on the water.

- Watch pressure gauge fall.

- Listen to the pump kick on.

- Turn off the water.

- Watch pressure gauge as pressure returns.

Personal Note

"One of my inspectors had judged a private water system to have good pressure and flow and the well equipment in good operation. But then he was asked to let the water run in order to pull a clean sample for testing. After the water ran an hour, the well went dry. When questioned, the owner said, 'Oh, yeah, the well often goes dry.' The customer was not happy to hear that.

"You are not required to report on the quantity of water supply and should let your customers know that testing the well equipment for 10 to 20 minutes does not confirm an adequate supply."

Roy Newcomer

NOTE: The home inspector should always suggest that the customer **have well water tested** for bacterial contamination by a reliable laboratory. Testing the quality of well water is outside the scope of the home inspector. Customers can also have a **yield test** done by a plumber to test the pumping capacity of the well. The home inspector may want to suggest this as a means of testing whether the water volume available will be appropriate for the customer.

It's necessary to test the well equipment in order to properly inspect it. Follow these steps to do so:

1. **Run water:** Turn on a faucet, perhaps at a laundry tub in the basement. Let the water run.

2. **Watch the pressure gauge:** As the water is running, locate the pressure gauge and watch the pressure drop until the pump is activated by the low limit. You'll be able to tell if the pressure gauge is working. And you'll be able to note other problems if the pump is *not* activated when pressure reaches the low limit.

3. **Listen to the pump:** As the pump is running, listen to it and note any problems with it. For the submersible pump, put the tip of your screwdriver against the pressure tank and your ear against the other end of the screwdriver. Excessive noise or vibration can indicate problems, which will be discussed below. Note also if the pump is turning on and off several times each minute.

4. **Turn the water off:** Turn off the faucet that's been running or have the customer do it so you can continue to watch the pressure gauge.

5. **Watch the pressure gauge:** Again, watch the pressure gauge as pressure returns. See if the pump stops when pressure returns to the high limit. If the pump doesn't stop or if it stops after bringing the tank past the high pressure limit, problems are indicated.

During the inspection of the well equipment, the home inspector should watch for the following conditions:

- **Inoperable pump:** When you conduct your test with the well equipment, the pump may not kick in as the pressure falls to or below the low limit. Be sure to report if the pump doesn't work at all. It's possible that there is no power to the pump caused by a blown fuse, poor electrical connections, or the pressure switch is shut off. If you can't determine why the pump has no power, then an electrician should be called to explore the situation. The pump could also not work due to the motor being burned out or bearings shot, a faulty pressure switch, or a seized or frozen pump. A well technician should be called to deal with those problems.

- **Constantly running pump:** If the well pump is constantly running before and/or during the test you perform, either the pressure tank is waterlogged, there may be a faulty pressure switch, or a leak in the pump piping. With the jet pump, the pump may have lost its prime. Be sure to report this situation. Feel the side of the tank for a marked temperature difference, indicating how much water is in the tank. You can often tell in this way if the tank is short on air and is waterlogged. A waterlogged condition can cause excessive wear on the pump and pressure switch, shortening the life of the well system. A waterlogged tank can be fixed easily by pumping more air into the tank.

- **Excessive noise or vibration in pump:** Listen to the pump equipment as the pump runs. Jet pumps that squeak and squeal probably have a defective motor bearing. With submersible pumps, listen to the pump with your screwdriver against the tank. It should run smoothly and quietly. If it sounds noisy, the bearings are probably going out. This should be reported and the suggestion should be made to have a well technician take a look at the pump.

- **Pump equipment loose or leaking:** Inspect any motor and pump assemblies above ground. They should be solidly secured on the bottom and not leaking.

- **Rusted, corroded, or leaking pressure tank:** Examine the pressure tank for any evidence of rusting, corrosion, or leaking. Sometimes, it isn't easy to tell the difference between water stains from condensation and leaking. Always check the bottom of the pressure tank to see if it's leaking and/or rusting out.

INSPECTING WELL EQUIPMENT

- Inoperable, noisy, vibrating, or constantly running pump

- Loose or leaking pump equipment

- Rusted, corroded, or leaking pressure tank

- Tank under or over pressure limit

- Missing or inoperable pressure gauge

- Missing, damaged, or leaking main shut-off valve

- Pipe too small

- Problems with well pit

Photo #5 shows a *pressure tank* in the corner of a basement. Notice the water and stains on the floor around the tank. This tank was leaking and was rusting out at the bottom. The dark water stain on the side of the tank is from leaking at the pressure gauge connection. A leak from a connection can be repaired, but a tank that is rusted out and leaking should be replaced. We suggested that this leaking tank be replaced and the new tank be insulated to prevent condensation from damaging the tank.

#5 Pressure tank

- **Under or over pressure limit:** If the well system is **below the low pressure limit** or does not come up to this limit while you're performing your test, there may be a faulty pressure switch, the pump piping may be leaking, or the well casing could be cracked. With the jet pump, the jet could be clogged. After your test, if the pressure comes up **over the high pressure limit**, then there could be a faulty pressure switch. The pressure tank may have a **relief valve** for excessive pressure buildup. The situation of the well system being under or over the pressure limits should be reported and a suggestion made to have the problem fixed.

- **Missing or inoperable pressure gauge:** Check to see if a pressure gauge is present and report it if missing. Often, these gauges no longer work. This should be reported.

- **Missing, damaged, or leaking main shut-off valve:** There should be a main shut-off valve between the entry of the well piping and the pressure tank. Point out its location to your customer and suggest that any damage or leaking should be fixed.

- **Do not operate the main shut-off valve.** You may only find that you can't turn it back on or the valve could start to leak after you turn it back on.

- **Incoming pipe too small:** Plastic piping is used quite often with private water supply systems. Plastic pipe from the well should be 3/4", as should copper pipe, to provide an adequate flow of water.

- **Problems with well pit:** The home inspector should inspect the pump equipment located in a well pit. Inspect the pit too. Well pits should be locked so children cannot fall into them. Report any unlocked or otherwise unprotected well pits as a **safety hazard**. Check out the hatch cover and report if it is rotted out. Take a look at drainage in the well pit, which should be good enough to protect the pump equipment in the pit. Often, there is a sump pump in the pit to pump out ground moisture that seeps in. Test this sump pump for operation as well as any other sumps found in the basement.

Reporting Your Findings

When you're inspecting the plumbing system, have your customer present. It's a smart practice. When the customer comes with you, you have an opportunity to fully explain the inspection and point out findings to the customer. Customer knowledge is a big step toward the prevention of complaint calls later.

With some of the home's mechanical systems, it's important for the home inspector to remind the customer *during the inspection* exactly what can and cannot be inspected. That's the case with the plumbing system. Help your customer understand that you are performing **a visual inspection** of the plumbing system and will not be able to inspect items hidden from your view — pipes behind walls, for example. Remind them of the **scope and the limits** of your inspection.

Be patient and take the time to discuss what you'll be inspecting and what you won't be inspecting. Customers may not understand that you can't pass judgment on the well casing or the submersible pump, although you'll be on the lookout for symptoms of problems. Remind them that you will not be testing the quality of water supply or adequate quantity and suggest the appropriate technicians who can do those things for them.

If you get calls several months later about new leaks or well problems, then you can remind the customer that you discussed

---

A NOTE

The home inspector is <u>not required</u> to actually inspect the well itself. Condition of the well casing, pump operation, water quality, and adequate water flow is best left to qualified technicians.

Personal Note

"One of my inspectors lifted off the lid of a well pit to find that a family of raccoons had moved into the pit. The mother raccoon was a bit touchy about a home inspector wanting to get into the pit. The inspector was not about to scrap with a mad raccoon and reported he was unable to inspect the well pit. The customer was very understanding."

Roy Newcomer

Report Available

The American Home Inspectors Training Institute offers both manual and computerized reports. These reports include an inspection agreement, complete reporting pages, and helpful customer information. If you're interested in purchasing the Home Inspection Report, please contact us at 1-800-441-9411

how you would not be able to inspect particular aspects of the plumbing system. Hopefully, the customer will remember that.

As you are performing the inspection of the water supply entrance, whether a public service or private well, be sure to explain the following patiently and in simple terms the customer can understand:

- **What you're inspecting** — the service pipe, the main shut-off valve, the pump motor, the pressure tank, and so on.

- **What you're looking for** — proper size piping, leaking, noisy well pump, rusted pressure tank, damaged shut-off valve.

- **What you're doing** — testing the service pipe to see if it's lead, running water to test pump operation, watching the pressure gauge, going into the well pit.

- **What you're finding** — a waterlogged pressure tank, missing main shut-off valve, leaking pump equipment, system that is under or over pressure limits, rotted well pit hatch, and so on.

- **Suggestions about dealing with the findings** — calling a well technician to investigate a pump with probable defective bearings, replacing a rusted pressure tank, having a plumber install or repair a faulty main shut-off, and so on. But with this caution — don't make uneducated guesses about how repairs should be made.

Filling in Your Report

Every home inspector needs an inspection report. A **written report** is the work product of the home inspection, and every home inspector is expected to deliver one to the customer after the inspection. Inspection reports vary a great deal in the industry with each home inspection company developing its own version. Some are considered to be excellent, while others are not very good at all. A workable and easy to use inspection report is important for a home inspector in terms of being able to fill it in. Of greater importance is its thoroughness, accuracy, and helpfulness to your customer. We can't tell you what type of inspection report to use, but let's hope that it's professional.

The **Don't Ever Miss** list is a reminder of those specific findings you should be sure to include in your inspection report. We list these items after years of experience performing home inspections. Missing them can result in complaint calls and lawsuits later. Here is an overview of what to report on during the inspection of the water supply entrance:

- **Entrance information:** First, identify the water service as public or private, writing this in the inspection report. Next, identify the type of piping. If you've found a lead service pipe, be sure to note that fact and suggest in writing that a water contamination test be performed by a qualified party.

- **Main shut-off:** Note whether or not a main shut-off exists on the service entrance pipe and report on defects such as damage and leaking. It's also a good idea to note the **location** of the main shut-off in the report. This is a helpful bit of information for the customer.

- **Well equipment:** If a service is private, include more information about the type of well pump present and where it's located. Note the presence of a pressure gauge and whether it's operating. If you haven't been able to test the gauge, it's best to report that fact. Record defects such as a waterlogged tank, rusted or leaking tank, and noisy pump conditions. If the pump bearings sound bad, you may want to include a suggestion about having a well technician examine the pump.

- **Major repair or replacement:** If you've found a well pump that is not operating, list it on the plumbing page of your report and on a summary page at the back as a major repair.

DON'T EVER MISS

- Lead service pipe
- Missing, damaged, or leaking main shut-off valve
- Inoperable, noisy, leaking, or constantly running pump
- Rusted, corroded, or leaking pressure tank
- Tank under or over pressure limit
- Missing or inoperable pressure gauge
- Unlocked well pit cover

OLDER PUMPS

For well pumps over 13 years old, always note them as <u>items requiring replacement within the next 5 years</u> in your inspection report. The pump will be reaching the end of its lifetime in 5 years.

WORKSHEET

Test yourself on the following questions.
Answers appear on page 25.

1. The home inspector is required to:

    A. Observe gas line piping.
    B. Observe water conditioning systems.
    C. Observe lawn sprinkler systems.
    D. Determine private water supply quantity and quality.

2. If the service pipe from a public source is galvanized steel, then:

    A. The pipe will not be magnetic.
    B. A scratch test will leave a dull scratch.
    C. A scratch test will leave a shiny scratch.
    D. The pipe will have lots of bends in it.

3. What material is the service pipe in Photo #1 at the back of this guide made of?

    A. Copper
    B. Galvanized steel
    C. Lead
    D. Plastic

4. The jet pump shown here is for what kind of well?

    A. Shallow well
    B. Deep well

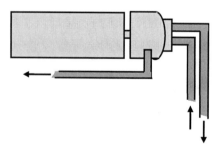

5. Which abbreviation is used when measuring water pressure?

    A. Gpm
    B. PB
    C. Psi

6. Which well component automatically starts the pump when pressure in the system falls?

    A. The pressure switch
    B. The pressure gauge
    C. The pressure tank
    D. The main shut-off valve

7. If, during a test of the well equipment, the pressure comes up over the preset high limit, what might be the problem?

    A. The bearings are shot.
    B. The well casing is leaking.
    C. The jet is clogged.
    D. The pressure switch is faulty.

8. What should be reported as a safety hazard?

    A. A corroded pressure tank
    B. An unlocked well pit
    C. A leak at the pressure gauge connection
    D. Plastic pipe used with a private well

9. What does it mean for a pressure tank to be waterlogged?

    A. There's water in the tank.
    B. The tank is leaking.
    C. There's too much water in the tank.
    D. The tank is full of air.

10. What is a symptom of a waterlogged pressure tank?

    A. The pressure gauge doesn't work.
    B. The tank returns to the high limit too fast.
    C. The pump vibrates too much.
    D. The pump runs constantly.

11. The home inspector is required to operate the main shut-off valve.

    A. True
    B. False

Chapter Three

WATER SUPPLY PIPING

The water supply system is the complete distribution piping system from the source of the water supply to all of the home's fixtures and faucets. The system includes **house mains**, which are the principal pipes to which all branches are connected. **Risers** carry hot and cold water vertically through one or more stories of the house. **Branch lines** carry water to the fixtures and faucets. Shorter lengths of rigid or flexible tubing connect piping to the fixtures and faucets.

The inspection of supply piping includes the inspection of the following aspects of the system:

- The **type and size** of piping material
- The distribution piping **installation** — locations and supports
- The **condition** of the pipes and connections
- The presence of **temporary** and/or **handyman plumbing**
- Functional **water flow**
- Branch lines to **outdoor faucets**

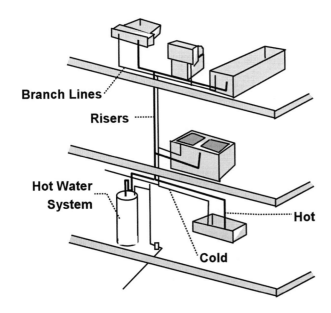

Corrosion

Before we discuss supply piping, let's talk a bit about what corrosion is. Corrosion is the physical or chemical change by which a material is degraded. In piping, corrosion leads to alterations in physical form and strength of the piping material.

INSPECTING WATER SUPPLY

- Piping material
- Installation
- Piping condition
- Temporary or handyman plumbing
- Functional water flow
- Outdoor faucets

Guide Note

Pages 23 to 40 present the study and inspection of the water supply distribution piping.

DISSIMILAR METALS GALVANIC SERIES

- Magnesium
- Zinc
- Aluminum
- Cadmium
- Steel
- Lead
- Tin
- Nickel
- Brass
- Bronze
- Copper
- Stainless steel
- Silver
- Gold
- Platinum

Definitions

Electrolytic or galvanic corrosion occurs when 2 dissimilar metals are connected to each other in water containing dissolved salts. The process releases metal ions, causing a current to flow.

Chemical corrosion occurs when metals react with oxygen, carbon dioxide, or salts in water. The process uses metal atoms to form new compounds.

There are 2 types of corrosion that affect plumbing pipes.

- **Electrolytic or galvanic corrosion:** This process takes place when there are **2 dissimilar metals connected to each other in water** containing dissolved salts, creating charged ions and tiny electric currents. The currents pick up ions from the metals, eventually destroying the pipes with leaks and failures in the walls of the pipes.

Galvanic corrosion often happens **from the inside** at the joints where dissimilar metals meet. An example would be at a joint with a galvanized steel plug in a brass fitting. It also happens **on the outside** of the pipes where dissimilar metals meet and there is water present from condensation. An example would be where copper pipes are hung with steel hangers in a damp place.

The important thing to remember is that it takes 2 dissimilar metals touching each other *plus* water or dampness for galvanic corrosion to take place. For this reason, supply piping should not have dissimilar metals in metal-to-metal contact. Professional plumbers use special couplings such as plastic, dielectric connectors, or tapes to prevent the flow of current across a joint. Amateurs tend to join copper to galvanized steel piping without any protection.

The box on the left lists dissimilar metals. The metals higher on the list will ionize first. Sometimes, dissimilar metals are used together on purpose so that one metal will be sacrificed first. An example is a magnesium or aluminum rod in a steel water heater that will give up ions first and thereby protects the steel.

- **Chemical corrosion:** This process depends on the presence of oxygen, carbon dioxide, or salts in the water that react with the metal. A chemical reaction takes place, using up some of the metal particles. The product of the chemical reaction either becomes soluble and washes away or it builds up in or on the pipes. An example would be rusting in cast iron pipes — the iron reacts with oxygen in the water to become iron oxide, or rust. Another example is what happens in lead pipes where lead salts are released into the water supply.

Types of Supply Piping

Water supply piping is generally larger at the supply end and smaller at the faucet end of the system. The supply piping is typically 3/4" ID for the cold and hot water mains in the house. Most branch lines use 1/2" ID piping.

The home inspector will find the following types of supply distribution piping used in homes:

- **Brass:** Brass is an alloy of copper and zinc. It was used as supply piping in some better homes in the early 1900's and in areas where water quality was bad for galvanized steel pipes, but not extensively. It hasn't been used in new construction for over 45 years. Brass pipes are threaded. Brass does not attract a magnet.

  There were 2 kinds of brass pipes used. **Yellow brass** is 67% copper and 33% zinc and lasts only 20 to 40 years, being more vulnerable to corrosion. **Red brass** (actually a yellowish-brownish color) is 85% copper and 15% zinc and lasts longer — generally 50 to 60 years but for as long as 75 years. Any brass pipes found today are likely to be near the end of their useful lifetime.

  Brass pipes can get **pinhole leaks** as they age. This happens when chemical components in the water cause the zinc to dissolve from the brass leaving tiny holes. Water can drip from the pipe or zinc salts can be left as the water evaporates. These **whitish mineral deposits** can self-heal the pinholes. However, the holes underneath the deposits will continue to grow. When the home inspector sees brass pipes with mineral deposits along their length, the customer should be told that the pipes have to be replaced.

  Similar signs of corrosion may be spotted at the threads, which can become paper thin. The joints can become so weak that applying a wrench can rupture the joint. If you see encrusted joints in brass piping, **don't lean against or push at them** — some are so weak that any pressure can rupture them. Any sections like this must be replaced.

- **Lead:** In rare cases, the home inspector may still find lead pipes that were installed over 80 years ago. Lead was once available for supply piping, although for safety reasons its use has long been prohibited. (It was also used in drain and waste piping, but that was prohibited in 1989.) Lead

Worksheet Answers (page 22)

1.	A
2.	B
3.	C
4.	B
5.	C
6.	A
7.	D
8.	B
9.	C
10.	D
11.	B

For Beginning Inspectors

*Inspect the visible supply
distribution piping in your own
home. Trace pipes in the
basement from the service pipe
and water heater and watch
where risers branch up to the
house. Identify the types of pipes
used. If you find dissimilar
metals used, check the joints and
look for galvanic corrosion.*

pipes are gray in color and can be identified by their distinctive bulb shaped joints. These joints were made by a process called **wiping**, where 2 lengths of pipe were inserted into each other and molten lead was smeared around the junction, forming the bulb. Lead pipes are not magnetic.

Chemical corrosion causes lead salts to be released into the water supply. Corroded areas with white deposits of lead salts are very brittle. Watch out! Touching them can cause even further damage to the pipes. The home inspector should **always report the presence of lead pipes** in the distribution system with the recommendation to have the water tested and to have these pipes replaced.

- **Galvanized steel:** These pipes are steel coated inside and out with corrosion-resistant zinc. The home inspector will find galvanized steel pipes used for supply piping in most homes built up to around 1960. These pipes are gray in color. Joints are threaded, and pipe dope, pastes, or Teflon tape were used to lubricate the joint when it was tightened. Galvanized steel pipes are magnetic.

Over time, the zinc coating is lost due to galvanic corrosion, then the steel pipe begins to rust from the inside. Rust can choke up the pipes, reducing water pressure. Eventually the pipes begin to leak when rusting becomes excessive, usually at the joints first. Exposed steel will also rust on the outside of the pipe due to water on the pipe or condensation. Galvanized steel pipes generally last about 40 to 60 years.

- **Copper:** Copper is the most common supply piping material used today and has been since about 1950. Copper pipes have varying wall thicknesses:

  — **Type K,** the thickest, is used underground.
  — **Type L,** medium thickness, is preferred for supply distribution piping.
  — **Type M,** lighter and thinner, is also used for indoor piping, although some communities require type L. Type M can fail in 10 years if the water is too acidic.

Copper pipes are too thin to be threaded. Instead, they are soldered at the joints. With **soldering**, the end of the copper pipe and the interior of the fittings are covered with flux, which eliminates chemical contamination. Then the elements are heated and a soft solder applied. These joints may be broken by mechanical damage from stress or poor supports. Since 1988, the use of lead solder is prohibited. It was found that lead can leach into the water supply from soldered joints in copper piping. When lead solder was used, the water supply should be tested.

Copper pipes may have joints that are **brazed**, where so-called silver solder is applied. These joints are stronger than those with soft solder.

Copper can take on a greenish cast, particularly around the fittings. This is usually the result of excess flux left during the soldering process. Leaks can cause a greenish corrosion of copper pipes. In some cases, where the water is highly acidic, copper may experience galvanic corrosion, releasing copper ions into the water supply. A sign of this is greenish stains in sinks and other fixtures. If these signs are found, the water should be tested for the presence of copper, which is a health hazard.

- **Plastic piping:** Plastic piping has been around since the 1970's. Codes vary in the use of plastic piping in the home. In some areas, it may be forbidden entirely; while in others, there may be requirements about using it for cold water supply lines only. In some places, it's acceptable for hot water supply lines too. In general, amateur and handyman plumbers are more likely to use plastic piping than professional plumbers. The presence of certain plastic piping can be a sign of other amateur work in the house.

Plastic piping needs more supports than metal piping. It must be supported every 4' or so to prevent sagging between hangers. Plastic piping should not be used within 6" of water heater fittings and should not run too close to recessed lighting fixtures and heat ducts.

- **PB (polybutylene)** piping is blue-gray in color and uses press-on fittings. It's fairly flexible and less likely than copper piping to burst under freezing conditions. But PB has lower high temperature limits than CPVC, so care must be taken not to get it too close to heat ducts. PB piping has had a history of failure, particularly failed fittings.

← Recalled

*"I wish I could say that I've
never caused a plumbing
problem during an inspection.
But it just ain't so.*

*"It only takes a split second to
get curious and take a poke at a
corroded joint, causing a major
break or leak in the system. It's
your job to point out problems,
not cause them.*

*"Paying for a plumber to come
in to correct the damage will
teach you not to do it again."*

Roy Newcomer

• **CPVC (chlorinated polyvinyl chloride)** is a cream or beige plastic piping that is solvent welded (glued) at the joints. CPVC is more rigid than PB and will normally split if frozen. It can stand higher temperatures than PB.

Other plastic piping is available. White **PVC** (polyvinyl chloride) and black **ABS** (acrylonitrile butadiene styrene) are generally used for drain and waste piping. Clear, red or blue **PEX** (cross-linked polyethylene) is the most recent type to hit the US market. It has been used in Europe for over 25 years.

The following chart summarizes the important points made about various types of supply piping.

Piping	Color	Joints	Corrosion
Copper	Copper turning brown with age	Soldered, brazed	Green corrosion on leaking pipes or of flux at joints
Red brass Yellow brass	Brownish Yellowish	Threaded Threaded	Loses zinc and gets pinhole leaks with white salt deposits
Galvanized steel	Gray	Threaded	Loses zinc and steel rusts out from inside
Lead	Gray	Wiped	Lead salts released into water supply
PB plastic	Blue-gra	Press-on fittings	Not applicable
CPVC plastic	White or beige	Solvent welded	
PEX	Clear, red, or blue	Press-on or expanded	

For Your Information

*You should be aware of local
codes on plumbing practices in
your area. Find and visit with a
plumber who can fill you in on
what the requirements are. Ask
questions about the public water
service and other concerns.*

Inspecting Supply Piping

During the inspection of water supply piping, the home inspector should keep an eye out for the following conditions:

- **Presence of lead pipes:** Always be on the lookout for lead supply piping and be sure to report it to the customer. Inform them that the water supply should definitely be tested and that pipes will probably have to be replaced.

- **Corroded or rusted pipes:** Look for corrosion in the piping and at the joints. Watch for mineral deposits on brass and lead pipes, rusting on galvanized steel, and greenish deposits on copper. (Refer to pages 23 to 28 for corrosion on various kinds of pipes.) Pipes that are corroded are nearing the end of their useful lifetime and will begin leaking in the future. Customers should be told that these pipes will eventually need replacement.

Be especially alert when you find **dissimilar metals** in lengths of piping such as the use of galvanized steel piping with copper piping. Often, deteriorated lengths of galvanized steel piping are replaced with copper piping, and the proper connections may not be used.

#6 Galvanized steel and copper pipe

*Photo #6 shows **galvanized steel and copper pipe** used in the same run. Here, we found a fitting deteriorating due to galvanic corrosion — notice the white mineral deposits on the fitting. This joint will begin leaking in the future. We suggested the fitting be replaced with a proper dielectric or plastic fitting to prevent further corrosion. This photo also shows a good view of solder joints in copper plumbing (at the right). Notice also what this homeowner has done with the floor joist — a huge chunk is cut out to provide room for the plumbing. Not a good idea.*

#7 Corroded pipe and valve

*Photo #7 shows a **corroded pipe and valve**. We did not turn off that valve or even touch it. This valve wasn't used on a regular basis and shows evidence of corrosion. If we'd turned it off and back on, it would have leaked for sure. Don't test valves in supply piping even if they look good.*

- **Drips and leaks:** Watch for dripping and leaking from visible piping and for water stains on ceilings, walls, and floors that indicate leaking. Try to find out where leaks are coming from when you see water stains. The source could be the pipes, at their connections to fixtures, or the fixtures themselves. Be sure to report any leaks you find. It's a high liability issue for home inspectors.

#7 Evidence of leaking

*Photo #8 shows **evidence of leaking** in the basement ceiling. In this home, there was evidence of leaking in the flooring above and in the joists. The pipes were showing corrosion from the water dripping on them. In general, it was not a pretty sight.*

Notice also the amount of electrical work in this same area — there are wires all over the place. In this case, we recommended that a plumber be called in to evaluate this situation. There was too much going on to tell what the source of the problem was.

- **Noisy pipes:** Pipes that are not very secure within the walls and not properly supported with hangers can vibrate, rattle, and bang as water moves through them. When hangers are too tight, pipes rubbing against them can cause a squeaking sound. Whistling and chattering can be a sign of loose or worn out washers in the faucets.

- **Water hammer** happens when water flow is cut off suddenly and a shock wave is sent back toward the supply end. It happens because flowing water has momentum. When the faucet is turned off quickly, the water slams into the valve with great pressure, creating a vacuum as it bounces back. The vacuum pulls the water against the valve again, repeating the process in smaller and smaller shock waves until they finally stop. These shock waves cause loose pipes to move and make banging noises. Over time, the repeated shock waves can damage connections. The following suggestions can be made to customers for the cure of water hammer:

  — Turning faucets off slowly
  — Tightening up supports for loose pipes
  — Installing an additional pipe, called an air chamber, past the take-off for the faucet. When water is cut off quickly, the momentum of the water will be carried into the air chamber and not backward into the pipes.

- **Improper installation:** All piping must be properly supported along the plumbing run. Each size and type of pipe has its own requirements for the type of hangers used and where supports should be. Watch out for the use of steel hangers with copper pipes and the resulting galvanic corrosion that can occur. Hangers that are not installed at proper distances or are loose or broken can damage piping.

  Hot and cold supply pipes should be 6" apart from each other (only 3" if insulated). Plastic pipes must meet requirements for distance from heat ducts and approved distances on lines to and from the water heater.

- **Insulation problems:** In certain circumstances, pipes should be insulated. Pipes in unheated areas like crawl spaces, basements, and garages should be insulated in cold climates to prevent them from freezing and bursting. In some homes, where cold water pipes sweat extensively, they should be insulated to prevent them from sweating and damaging the pipes from condensation.

You may find pipes already insulated with materials such

CAUTION

Do not operate any valves in supply piping. It's far too common for unused valves to start leaking, and it may be impossible to turn them back on. You don't want to be the cause of plumbing problems.

Personal Note

"One of my inspectors was inspecting a home for an old man who was just recovering from heart surgery. The customer wanted the inspector to see if the bypass valve to the water softener system worked.

"The inspector said no, but the customer virtually begged him to try it. Well, the inspector is a nice guy and couldn't refuse this sick old man, so he did.

"Water started shooting out, and it wouldn't stop when the valve was turned off. Both the inspector and the old man were soaked to the skin. We paid to have a plumber repair the valve. I suppose we were fortunate not to have to pay hospital bills too."

Roy Newcomer

as asbestos, oakum, felt, or sphagnum moss in canvas jackets or newer materials such as mineral wool and glass fibers or plastic foam. **Asbestos insulation** poses a danger to people when it is disturbed and asbestos particles are released into the air and breathed in. The home inspector isn't trained to determine whether piping insulation is asbestos or not. But if asbestos insulation is suspected, the inspector should indicate in the report that an asbestos type product is used in the insulation.

*Photo #9 shows **plumbing pipes in a crawl space**. We found this crawl space in Wisconsin, where winters are pretty bad. These pipes should be insulated for 2 reasons. First, the crawl space is unheated and un-insulated, and the pipes should be insulated to protect them from freezing. Second, there's no vapor barrier in this crawl space, and the high humidity had been causing condensation problems. Here, you can see that the large waste pipe shows signs of corrosion. We recommended insulating the pipes.*

#9 Plumbing pipes in a crawl space

- **Missing valves:** In the supply distribution system, there should be a number of turn-off valves to allow work to be done on the system. There should be a turn-off valve on the cold water line to the water heater. Most toilets have their own turn-off valve, and they're often present under every sink. We find with the newest construction that the turn-off valves are not being put in due to the cost. Plumbers are apparently asking about each cost and giving homeowners a choice. Missing valves are not a serious problem and don't have to be written up in the report, but it is a good idea to point out their absence to the customer during the inspection. Don't forget — don't operate these valves.

- **Inoperable hose bibs:** Hose bibs are outdoor faucets. The home inspector **should test hose bibs for operation** except in the winter in northern climates. The **conventional hose bib** setup consists of a regular faucet outside with a hose bib shut-off valve inside the basement. In winter, the valve should be closed and drained and the faucet left open to prevent freezing. A special **frost-proof hose bib** is available with a long stem that turns off the water supply inside the house each time the faucet is used. The frost-proof hose bib does not need to be shut off during the winter.

Record whether hose bibs are present outside the house and test each one for operation. Do *not* test hose bibs in the winter in cold climates. The home inspector should report whether hose bibs were tested or not.

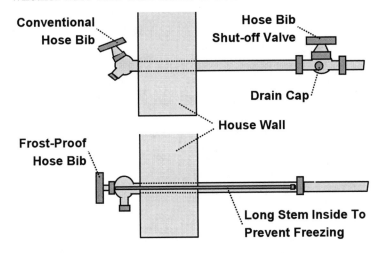

- **Temporary repairs:** Watch for and report any repairs to the plumbing system that are temporary. Some homeowners may use rubber tubing to replace a section of piping. The use of rags, tape, putty, or wood plugs to repair leaks should be reported and fixed professionally.

#10 Interesting temporary repair

*Photo #10 shows an **interesting temporary repair**. The plumbing shows evidence of serious corrosion from leaking. Look closely at the left side of the photo. A file has been stuck into a hole in the pipe to plug a leak — it's sticking perpendicularly out of the pipe. We carefully avoided touching the file, because to pull it or knock it out of position would have caused the pipe to leak again. We recommended that a plumber come in to repair the pipe.*

- **Handyman plumbing:** Inspect the supply lines closely for evidence of handyman plumbing. Signs of handyman work may be the use of plastic pipes, poorly made connections, using dissimilar metals next to each other without proper fittings, and inadequate supports. Report any handyman work you find.

#11 Handyman plumbing

*Photo #11 shows **handyman plumbing**. We didn't assess whether the plumbing worked, and it may have. But we did feel that the work being done in such close proximity to the electrical panel was not a safe situation. Here, the electrical panel is missing a cover and actually couldn't be covered. If this handyman plumbing job starts to leak, you've got a serious safety hazard. We recommended that a plumber be called in to check this out and decide how to reroute this plumbing line.*

Personal Note

"One of my inspectors came upon a home with handyman plumbing. In this house, the homeowner had used flexible copper piping to replace corroded pipes. The flexible copper piping went every which way.

"The inspector had young children and must have spent a lot of time reading to them, because he said the plumbing looked like a drawing in a Dr. Seuss book."

Roy Newcomer

Functional Flow

Supply piping should also be tested for functional flow during the inspection. The home inspector should be able to report if the functional flow is satisfactory or poor.

There are 2 aspects to water flow. The first is **adequate flow**, whether or not there is enough water volume flowing. The second is **water pressure**, that is, whether or not there is enough push behind the flow. In general, these 2 aspects of water flow mean the same thing to homeowners. It's true that when water pressure is low, the water flow is low. But when the water flow is low, it doesn't necessarily mean low water pressure. There can be other problems with the plumbing system.

The question for the purpose of the home inspection is whether the water at faucets is **functional**. Is there enough water coming out of the faucet fast enough to be functional? We all know when the flow isn't functional. That's when water trickles out and takes a long time to fill a sink or tub.

City water systems usually deliver water to homes at a pressure between 40 and 70 psi. Private wells may maintain a water pressure anywhere between 40 and 60 psi, depending on the pressure switch settings. When no water is running in the house, the water sits in the plumbing at this static pressure. When a faucet is turned on, pressure drops as water moves through the pipes. It might drop 2 psi for every 10 feet of pipe, depending on several factors. Pressure also drops about 8 psi as it rises from the basement to the second floor of the house, due to the force of gravity acting on it. This is natural. Water pressure at the faucet is expected to be lower than the static pressure. When additional faucets are turned on, water pressure drops again. But taking these conditions into consideration, water flow should still be adequate at faucets.

The home inspector should perform the following tests and determine whether water flow is functional based on his or her opinion of what functional flow means:

1. Run the water at the kitchen sink and note whether water pressure is satisfactory.

2. In the bathroom, turn on both sink and tub faucets and then flush the toilet. You can expect water pressure to drop off some, but not entirely. Judge whether water flow is still functional. For example, if water to the tub stops completely, water pressure should not be reported as satisfactory.

3. Test other faucets in the home, garage, and outside as you did in the kitchen.

The home inspector does not have to find the cause of poor water flow in the home. However, it helps to be able to tell the customer what the possible causes are. A plumber should be called to determine the exact cause. Here are some possible causes of poor water flow:

- **Constricted pipes:** Water supply pipes can become constricted due to corrosion, clogging, or damage which crimps the pipes. Galvanized steel supply piping rusts out from the inside as it ages, reducing the diameter of the pipe. There may be an adequate flow of water under enough pressure, but a constricted pipe will move less water than normal.

- **Partially closed valve:** If the main shut-off valve or any other valve between the source and the faucet is partially closed, water flow will be reduced.

FUNCTIONAL FLOW

Water flow is functional when <u>enough water</u> comes out of a faucet <u>fast enough</u>. The home inspector determines whether flow or pressure is okay or not, based on his or her own opinion and common sense. Let your customers watch and decide if they agree.

Personal Note

"One of my inspectors turned the faucets on in the bathroom sink and plugged the sink to check flow and drainage. However, while the sink was filling up, he was distracted by the customer and left the bathroom to look at something in the hall.
"A few minutes later, he was shocked to see water flowing out of the bathroom into the hall. He'd forgotten about the sink.
"I don't have to tell you not to do that, do I? Keep your attention focused on what you're doing at all times."

Roy Newcomer

- **Wrong sized pipe:** If a section of pipe with a smaller diameter than is required is inserted in the water supply piping, the water flow will be reduced.

- **Too many changes in directions of the pipes:** Excessive bends and corners in the piping slows the water flow down.

- **Low water volume:** A flow rate of 5 to 7 gpm is considered to be adequate in most parts of the country. A peak flow rate of 10 gpm is the optimum flow rate for a modern home with 2 bathrooms. It's possible for a public water service to provide water at an unacceptable flow rate to service the house. The problem could be a clogged main to the house. With private systems, the well may not have enough yield to provide enough water to the house or there could be a problem with the pump itself.

- **Low water pressure:** The city may provide water at a pressure too low for functional use in the home. The installation of a booster pump and pressure tank in the home may solve this problem. With private wells, the pressure switch could be broken or out of adjustment.

Cross Connections

The last deficiency in the water supply piping system to look for during the inspection is the presence of cross connections between water supply piping and the drain and waste piping. A cross connection exists where water from the drain and waste system can be siphoned back into the supply system. Waste being introduced into the water supply system can contaminate the water. Any cross connection presents a **safety hazard**.

A cross connection can occur when a faucet spout extends lower than the water level in a basin such as a laundry tub. If water pressure was lowered suddenly, the water in the basin could be sucked back into the supply pipe. The solution is to raise the faucet so the spout can't reach the water level of the basin. A cross connection could also occur if a hose is left in a bucket of water outside. Again, a lowering water pressure can siphon water from the bucket back into the water supply system.

Cross connections at some equipment and fixtures are prevented through the use of **anti-siphon devices** and **vacuum breakers**. Faucets can also be fitted with anti-siphon devices.

The home inspector is not required to determine the effectiveness of these anti-siphon devices.

During the inspection of the water supply piping, the home inspector should keep an eye out for any possibility of cross connections and to report any found as a safety hazard.

Reporting Your Findings

Depending how your inspection report is laid out, you may be reporting your findings from the inspection of water supply piping on a general plumbing page or on several pages. Supply piping is checked in the kitchen and bathroom as well as wherever it's visible in the basement or other areas in the home.

- **Kitchen:** After you test for **functional flow** of water at the kitchen sink, rate the flow as adequate or poor and record that fact in your inspection report. (You should report this on the kitchen page of your report.) Sometimes, it's better to talk about functional flow as water pressure. That's a term customer's will understand.

- **Bathroom:** Rate the functional flow of water in the bathroom as suggested above. Note that for now, we're only concerned with the water supply. Drain piping, faucets, toilets, and so on are presented later in this guide.

- **Supply piping:** On the plumbing page of your inspection report, identify the supply distribution piping material used in the home. If you've found lead supply pipes, be sure to highlight that information and suggest that a water contamination test be performed by a professional. Note in the report that lead supply piping will probably have to be replaced. Record the defects found in the supply piping such as those listed here and suggest when a plumber should be called in for repair:

  — Corroded, rusting, damaged, noisy, and leaking pipes
  — Improperly supported or insulated pipes
  — Handyman or temporary fixes

- **Hose bibs:** Always note whether or not hose bibs are present, whether or not you've tested the hose bibs, and whether or not they're operating.

DON'T EVER MISS

- Presence of lead pipes
- Corroded or rusted pipes
- Leaks, old or new
- Inoperable hose bibs
- Handyman plumbing
- Temporary plumbing
- Poor functional flow
- Cross connections

- **Water stains and leaking:** Report water stains on ceilings, walls, and floors wherever you find them. The source of the leak may not be supply piping, but you'll want to be sure not to miss them. It's a good idea to report leaks on the page of the report dealing with the location — for example, kitchen, bathroom, bedroom pages or other locations in the home.

- **Safety hazards:** Never miss reporting any safety hazards you've found. For the plumbing system, a **cross connection** represents a safety hazard. Be sure to report a cross connection on your plumbing page and then repeat it on a summary page at the back of your report. It's always a good idea to list all safety hazards found in the home, not only in the plumbing system, on a summary page.

- **Major repairs:** Take it easy on defining plumbing problems as major repairs. Some repairs are minor and only require a few hours of a plumber's time to fix. Save the category of major repairs for truly serious plumbing problems.

WORKSHEET

Test yourself on the following questions.
Answers appear on page 40

1. Which type of supply pipe will rust from the inside out?

   A. Red brass
   B. Copper
   C. Galvanized steel
   D. Lead

2. Which type of supply pipe is soldered?

   A. Yellow brass
   B. CPVC
   C. Copper
   D. Galvanized steel

3. Which type of supply pipe is especially subject to pinhole leaks?

   A. Brass
   B. CPVC
   C. Galvanized steel
   D. PB

4. Rusting is an example of:

   A. Chemical corrosion.
   B. Electrolytic corrosion.

5. What is electrolytic corrosion?

   A. A reaction of metal atoms with oxygen, carbon dioxide, or salts in water
   B. A process where 2 dissimilar metals connected in water give up ions and make a current.
   C. A process called wiping that causes bulges in lead pipes.

6. Which of the following types of supply piping if least likely to burst in freezing temperatures?

   A. Copper
   B. PB
   C. CPVC

7. What is the purpose of an anti-siphon device?

   A. To provide a turn-off so the homeowner can work on a toilet or sink
   B. To prevent water hammer
   C. To prevent pipes from freezing
   D. To prevent waste water from backing up into the water supply system

8. What should be reported as a safety hazard?

   A. The use of CPVC piping
   B. A cross connection
   C. Greenish color at joints in copper piping
   D. Handyman plumbing

9. When should the home inspector suggest a water contamination test?

   A. When greenish stains are found in sinks
   B. When dissimilar metals are used in a single plumbing run
   C. When asbestos insulation is found on pipes
   D. When water pressure is low

10. Which of the following statements is true?

    A. The home inspector is not required to operate hose bibs.
    B. The home inspector should always operate hose bibs.
    C. The home inspector should operate hose bibs except when they are winterized.
    D. The home inspector should operate hose bibs only when they are winterized.

11. What would not be the cause of low water flow?

    A. Low water volume
    B. Constricted pipes
    C. Partially closed valves
    D. Missing valves

DWV INSPECTION

• Piping material

• Installation

• Piping condition

• Temporary or handyman work

• Functional drainage

Guide Note

Pages 40 to 59 present the study and inspection of the drain, waste, and venting system.

Worksheet Answers *(page 39)*

1.	C
2.	C
3.	A
4.	A
5.	B
6.	B
7.	D
8.	B
9.	A
10.	C
11.	D

Chapter Four

DRAIN, WASTE, AND VENT SYSTEM

The drain, waste, and venting (DWV) system consists of all the piping that carries water and waste from fixtures to the public sewer or private septic system. Any discharge pipes in the DWV system are called **drain pipes**. Some drain pipes carry rain water or foundation seepage. The drain pipes that carry water away from fixtures are called **waste pipes**. Waste pipes that carry waste from toilets are called **soil pipes**. Vertical pipes in the system are called **stacks**. **Vent pipes** serve to release gases and pressures that build up in the system. The vent system exhausts above the roof.

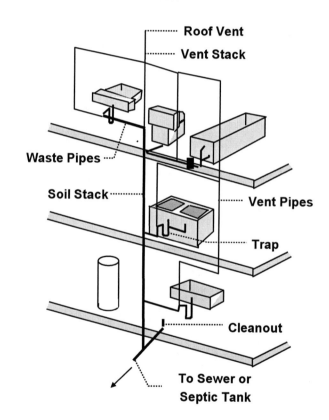

The inspection of the DWV system includes the inspection of the following aspects:

• The **type and size** of piping material
• The DWV piping **installation** — location and supports
• The **condition** of visible pipes, connections, and traps
• Functional **waste drainage**
• The presence of **temporary** or **handyman work**

Waste Disposal

A home may be connected to a public or private waste disposal system. **Public systems** in older neighborhoods may have combination waste and storm sewers, which when overloaded with storm water, can back up through the basement's floor drains. New neighborhoods have separate waste sewers and storm sewers. In this case, waste pipes and floor drains from the home exit into the city waste sewer, while runoff from gutters and downspouts exits through a drain tile system into the storm sewer.

A **private waste disposal system** is most often a septic tank and a seepage system, although cesspools were previously allowed and some may still be in use.

A **cesspool** is a hole in the ground lined with masonry which was laid to be porous. Liquids in the waste system passed directly through the masonry into the ground. Solid waste was held back by the masonry and was broken down or digested by bacteria. Oxygen is available in a cesspool to feed the **aerobic bacteria** which live in the presence of oxygen. Cesspools can either become clogged with too much solid waste if they're not porous enough or pass solid waste too easily into the soil if they're too porous. Most communities no longer allow cesspools.

A **septic tank** is an oxygen-free environment that depends on **anaerobic bacteria** to break down wastes. The tank can be made of concrete, steel, or fiberglass. The entering waste from the house separates with lighter materials rising and solid matter dropping to the bottom of the tank. Bacteria then change solids into simple chemicals that dissolve. Baffles prevent anything but liquid from being discharged from the tank.

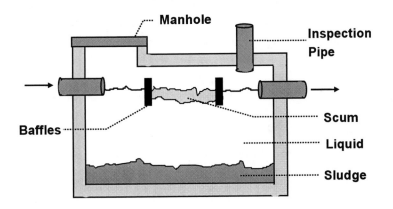

SEPTIC SYSTEMS

The home inspector does <u>not</u> have to inspect the septic system. It is beyond the scope of the general home inspection.

Definitions

A <u>cesspool</u> is a masonry lined hole used to hold and break down solid materials from the home's waste system before releasing them into the ground through the porous masonry.
A <u>septic tank</u> is a watertight underground tank of concrete, steel, or fiberglass into which household waste is held and broken down for release to a seepage field.
<u>Aerobic</u> bacteria thrive in an oxygen-rich environment, while <u>anaerobic</u> bacteria live in an oxygen-free environment. Both types of bacteria break down solid waste matter.

The liquid discharged from the septic tank, which is still being worked on by bacteria, flows to a distribution box and into one of many porous drain tiles in a **seepage or leach field or pit**. There the liquid is further purified as it makes its way down to the water table. The tank, distribution box, and tiles in the seepage field must meet local distance requirements from the house and the private well. These distances vary from community to community, but approximate those shown here.

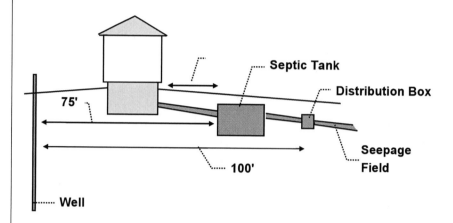

The inspection of the septic system is beyond the scope of the general home inspection, and the home inspector is **not required to inspect it**. The most a home inspector can be expected to check is whether correct distances are met for local codes and whether there is pooled water appearing above the seepage field or an odor coming from the field.

Septic tank contractors can be called in to exam the tank and the seepage field. Such an inspection would include running dye through the system and checking for leaks at the tank or seepage field and inspecting the interior of the tank, its contents, and the condition of all its components. The contractor may find problems such as an overloaded tank or field, broken or damaged components, clogged tiles, and soil breakdown at the seepage field. Such problems can cause a septic system to fail and require replacement.

If a private septic system is present on a property being inspected, the home inspector can inform the customer that the septic tank needs to be pumped out every few years. The inspector may also recommend that a septic tank specialist come in to inspect the tank. Most communities require this anyway when a home is sold.

For Your Information

Check your local codes for regulations regarding septic systems in your area. Local distances can vary from those given here. Check your codes for requirements regarding waste piping and venting as well.

The Venting System

The part of the DWV system in a house that is least understood is the venting system. Venting maintains all parts of the DWV system at atmospheric pressure so that gravity can clear the waste pipes. It also serves these 3 functions:

- It allows air in front of the water going through the pipes to be pushed out of the way.
- It allows air to be reintroduced into the piping after the water has passed.
- It allows sewer gases to escape from the system.

Vent piping from each fixture joins a **main vent or soil stack** above the waste line (see drawing on page 40). The main vent stack extends through the roof, terminating 6" or more above the roof. The roof vent must also be 1' away from any wall and 10' from any window, since sewer gases will be expelled through the vent. In ranch-style homes with a bathroom at one end and kitchen on the other, there will typically be 2 vent stacks exhausting through the roof.

Most vents are **dry vents**, meaning they carry air and water vapor only, not waste. A **wet vent** provides ventilation from a fixture on a floor below as well as carrying waste from a fixture on the floor above, combining venting and drainage. The wet vent should be a larger size, but this is often not done, especially when bathrooms are added during remodeling. Wet vents are not allowed in some communities. Vent piping is usually hidden in the walls.

Every fixture drain has a **trap** to hold water in the plumbing system and to prevent the backflow of gases from the venting system. The amount of water held by the trap is called the **water seal.**

The **P-trap** is commonly used today below plumbing fixtures and in basement floor drains where the trap is under the slab. The water seal is maintained at the level shown in the drawing here.

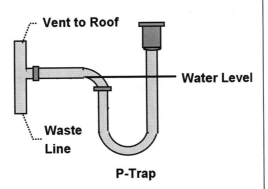

Definitions

A <u>dry vent</u> in a home's DWV system carries only air and water vapor. A <u>wet vent</u> combines both the venting of air and water vapor and the carrying of waste matter. A <u>trap</u> holds water in the plumbing system and prevents backflow of gases. Different kinds of traps include the <u>P-trap</u>, which is shaped like the letter P, the <u>S-trap</u>, shaped like the letter S, and the <u>drum trap</u>.

It isn't possible for the home inspector to know for sure if a P-trap is, in fact, vented. However, any drain that gurgles and belches as it drains is not at atmospheric pressure, which indicates the absence of proper venting or some obstruction in the vent piping.

The **S-trap**, although no longer permitted in modern plumbing work, was once used and can still be found. The S-trap is an unvented trap. They can lose their water seals during a discharge elsewhere on the stack, permitting gases to escape into the home. The best solution for old S-traps is to replace them with vented P-traps.

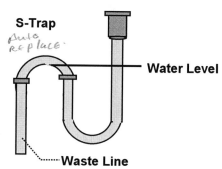

Photo #12 shows an air admittance valve, which may be installed when replacing S-traps with P-traps in an unvented location. The vertical black pipe between the 2 P-traps in this photo is the mechanical vent, which is not vented to the outside. This vent has a little gasket or flap that moves to form a seal and allow air into the pipe. The mechanism in this type of vent can fail, and for that reason, some communities do not allow their use. The home inspector should be familiar with local codes regarding their use.

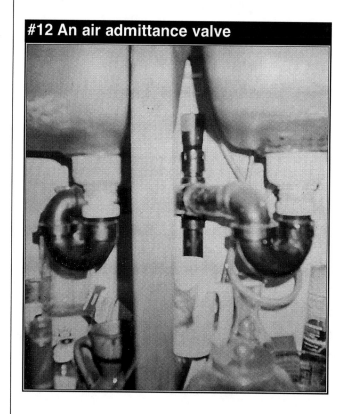

#12 An air admittance valve

Another type of trap that was used in the home is the **drum trap**, which can still be found with fixtures such as bathtubs and laundry sinks. The drum trap is not a vented trap either. The old drum

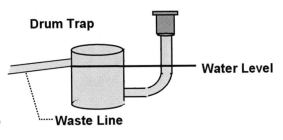

traps can leak and should be inspected carefully if they are visible. Often, the drum trap from a first-floor bathroom can be seen from the basement ceiling.

Amateur plumbers often ignore the venting system, and it is not unusual to find new fixtures put in without being connected to the venting systems. Amateurs often omit traps under sinks and other fixtures.

DWV Piping Materials

The size of DWV pipes used in the house varies, depending on their location and purpose. Common sizes are shown in the chart at the right. Older homes may have DWV piping that doesn't conform to this table. In general, lines carrying solid waste are larger than those carrying liquids. The home inspector will find the following types of DWV piping used in homes:

- **Cast iron:** Cast iron pipes in at least 3" diameters were commonly used for the waste and soil piping up to the 1960's, especially for the main stack and large horizontal runs. (It should be noted that in DWV piping the term *horizontal* is not entirely accurate — so-called horizontal piping must have a **slight slope** in order to allow wastes to flow downward.)

Cast iron piping is black, and is, of course, magnetic. A cast iron pipe is manufactured with a **hub** at one end. Pipes are joined by inserting the hubless end into a hub, packing with oakum, and filling with molten lead. Another method is the no-hub joint where hubless lengths of pipe are joined with **slip joints** with neoprene sleeves clamped on. Slip joints are used when cast iron is connected to copper, galvanized steel, or plastic piping.

DWV PIPES	
	Sizes
Building drain	4"
Soil stack	3" or 4"
Sinks	1 1/2"
Toilet	3" or 4"
Clothes washer	2"
Vent stacks	
No toilet	2"
Over toilet	3" or 4"

HORIZONTAL DWV PIPES

"Horizontal" pipes in the DWV system are not really horizontal. They must have a slight slope to them in order to carry away wastes.

Cast iron can last 50 years or more. Over time, it will rust from the inside. Rust can self-heal the pinholes for a time, but the pipes will eventually leak. Although DWV lines don't hold water all the time as supply lines do, horizontal runs of cast iron are vulnerable to corrosive sewer gas which eats through the top of the pipe first. Horizontal runs will corrode before the vertical runs. Impurities in the metal can cause rusting or even splitting lengthwise along a seam.

- **Galvanized steel:** Galvanized steel may be used in DWV piping for both drains and vents. In some areas, it's used only for venting where it lasts up to 50 years because it is not in contact with water. However, sewer gases can even eat at galvanized steel, leaving large holes. Galvanized steel has a shorter life expectancy when used for drain and waste lines. (See page 26 for more information.)

- **Chrome-plated brass:** Used almost exclusively for traps under sinks, chrome-plated brass is usually connected under the floor or behind the wall to other metal or plastic pipes. These traps do leak after a time. Watch for salt deposits on the underside of the traps. (See page 25 for more information.)

NOTE: It was quite common for homes to use cast iron stacks, galvanized steel smaller-diameter lines, and chrome-plated brass traps together. If you see cast iron stacks and chrome-plated traps, it's a good guess that galvanized DWV lines are hidden in the walls.

- **Copper:** The use of copper DWV lines is relatively rare, although it was used for a while from the 1940's to the 1960's. If used, they will be **type M**, the thinnest available. (See page 26 for more information.)

- **Lead:** Lead piping was used up to the 1950's to connect fixtures to the main stack or drain, especially for lead bends under toilets. Today, these lead pipes are likely to be leaking at the connections. These waste lines are likely to be replaced during any major plumbing work.

The lead used in DWV piping does not pose the same danger to people in the home as it does when lead supply piping is used. However, the use of lead even in DWV

For Beginning Inspectors

It's time to look at DWV piping in your own home. Of course, most of it will be hidden in the walls, but there is something to be seen in the unfinished basement, under the sinks, and in the attic. Start in the basement, locating the main stack and identifying the type of piping used. Trace lines into the basement ceiling and notice any signs of leaking. Identify the types of traps found under the sinks. Take a look in the attic to check that the vent stack is present and exhausts through the roof.

piping has been prohibited since 1989. (See page 25 for more information.)

- **Plastic:** For the last 30 years or so, rigid plastic piping has been used almost exclusively for DWV piping. **PVC** (polyvinyl chloride) piping is most common. It's either white or beige in color. Black **ABS** (acrylonitrile butadiene styrene) is also used.

  Both PVC and ABS are joined with **mechanical no-hub joints**. The drawbacks of plastic piping is that they are noisier than metal pipes and water can be heard rushing through them, they may burst if unprotected in freezing areas, and they throw off noxious fumes during a house fire.

The chart on the below summarizes information about DWV piping. We haven't repeated all the information about copper, galvanized steel, and lead piping that was presented earlier. See the chart on page 28 for more information on other metal piping also used in the DWV system.

Piping	Color	Joints	Corrosion
Cast iron	Black	Hub caulked with lead or no-hub with slip joints	Rusts, the horizontals rust out at top first.
Chrome-plated brass	Shiny silver	Threaded	Loses zinc and gets pinhole leaks with white salt deposits
ABS plastic PVC plastic	Black White or beige	Mechanical no-hub joints	Not applicable

Inspecting DWV Piping

During the inspection of the DWV piping, the home inspector should watch out for the following conditions:

- **Corroded, rusty, or leaking pipes:** Check all visible DWV piping for corrosion, rust, and leaks. Remember that horizontal runs of cast iron are likely to develop problems on the top of the pipe first. Check connections for corrosion, especially if 2 dissimilar metals are connected. Corrosion will stop leaks for a while, but eventually the leaks will start. The home inspector should report corroded and rusted pipes and let customers know that leaking will follow. The pipes will ultimately have to be replaced.

#13 Patched and leaking cast iron pipe

*Photo #13 shows a **patched and leaking cast iron pipe**. This large cast iron pipe had rusted along the top in the past because sewer gas had eaten through, and it had been patched. The pipe continued to deteriorate and now the bottom of the pipe is also leaking, right through the patch. Notice the rust on the galvanized pipe right next to it from this leaking. We recommended that the cast iron pipe be replaced.*

Take another look at **Photo #9**. This cast iron pipe in the crawl space also shows signs of corrosion along the top of the pipe. Look at **Photos #6 and #10** again for examples of corrosion in DWV piping.

When inspecting the DWV system **always look for water stains** on the walls, floors, and ceilings that can indicate a plumbing leak and record it in your inspection report. This is a **high liability** area. You may not always be able to tell the customer the cause of the leak, but you better point out the stain to the customer and then put it in writing to protect yourself against any later plumbing catastrophe.

#14 Leaking under the bathtub

Photo #14 *shows* **leaking under the bathtub**. *Watch out for signs of leaking at fixture drains. Here the bathtub drain is corroded, leaking and dripping water on the turn-off valve below, which is also corroded, and down the wall. By the way, we did not touch that valve. You just know that turning that valve would cause a whole new leak in the system. We did, however, point out the corrosion and leaking, warning the customer about the valve. In this photo, note the use of the drum trap under the bathtub.*

- **Improper installation:** When inspecting DWV lines, be sure to make note of horizontal lines. There should be a downward slope to these lines. Make sure the pipes are properly supported and hangers are in good condition so that lines run straight and no low spots develop in the lines. Waste can get hung up in the pipes if they sag.

- **Lack of insulation:** (See comments on page 31.)

- **Illegal or leaking traps:** Always check for traps at plumbing fixtures. Older **P-traps** under sinks are commonly rusted out and have loose joints, causing drips and leaks. Test the trap by gently grabbing hold of it and moving it *slightly* back and forth. But be careful not to be too rough with it. Rusted traps can fall apart if handled too roughly. Report the use of **mechanical vents** if local codes do not permit them.

Don't miss reporting **S-traps** if they're not permitted in your area. And let customers know that an S-trap isn't vented. Water can be sucked out of the trap, letting sewer gas into the house. If an odor of sewer gas is present, suggest that the trap be refilled with water to create the water seal. If it's a new house and the S-trap is plastic, it's most likely a sign of handyman plumbing.

Personal Note

"Try not to be too rough when testing traps. A gentle move back and forth will tell you if the joints are loose.
"One of my inspectors was inspecting a galvanized steel trap under a sink and gave it a vigorous shake. The result was not good. He actually crushed part of the trap, rendering it completely useless. We had to pay to have a new trap put in."
Roy Newcomer

Also, check visible **drum traps** for rusting and leaking and inform customers about their unvented condition.

When a home is served by a public water supply, there is probably a **house trap** where the home waste lines join the public sewer. The house trap is a large U-shaped fitting with capped heads at the top of each arm. There may also be a fresh air vent on the side of the trap. The sewer outlet fitting must have a **cleanout** sealed with a plug that can be removed if any work needs to be done on the sewer lines. This plug should be kept in place and sealed. Sometimes, the home inspector will find rubber patches, wooden bungs, or rags stuck in the cleanout. Improperly sealed cleanouts should be reported.

- **Missing or improper venting:** Most of vent piping is hidden from view, so the home inspector won't actually see too much of it. However, there are some things to look for. Unvented traps should be reported. Remember that S-traps and drum traps are not vented. Even P-traps can be installed and not connected to a venting system. A sign that a fixture is not vented is a gurgling or belching noise when the fixture drains. If you suspect that a P-trap is not vented, mention it to the customer and record it in your inspection report. Improperly installed vent pipes should be reported as a **safety hazard**.

If you notice the **odor of sewer gas** anywhere in the house, be sure to point it out to the customer and report it. Sewer gas can indicate missing venting. Be sure to check the roof to see if there are any exiting vent stacks.

Refer back to **Photo #5**. Notice the black drum trap under the laundry sinks. There's no venting from the sinks. If your local code requires laundry sinks to be vented, this should be pointed out to the customer. This may have to be corrected when the home is sold.

Be sure to check the **vent stack in the attic** and from the roof for proper extension and distances. The vent penetration should be checked for proper flashing and whether leaking is occurring. Look at the vent as it passes through the attic for corrosion. Leaking through the roof and high attic condensation can damage vent piping here.

#15 Attic vent

*Photo #15 shows an **attic vent** with large holes in it. In this case, you can tell that the roof has been leaking in the area of penetration, probably from inadequate flashing, and has caused the pipe to corrode badly. Sewer gas is escaping into the attic. Holes in vent pipes should be reported as a **safety hazard**.*

- **Temporary repairs:** Keep an eye out for repairs to the DWV line and to traps. The use of rags, tape, putty, and plugs in piping to repair leaks should be reported and fixed professionally.

- **Handyman plumbing:** New plastic S-traps and PVC or ABS pipes inserted into the plumbing runs are a signal to keep your eyes open for other amateur work in the system. Homeowners often put in plastic piping in the waste line but don't pay proper attention to supports. As a result the waste line may sag, causing waste to clog the system. Report any handyman work you find.

*Photo #16 shows an example of **handyman plumbing**. Here the owner put a run of plastic piping in a waste line, a sign of handyman work. Notice the black hose right below the new PVC piping. The owner has a hose going into the drain pipe. Now follow the drain pipe and see where it goes. It angles downward and runs into the soil stack. What's wrong with that, you ask? Well, there's no seal whatsoever where that pipe enters the soil stack, and there definitely should be. Sewer gas was pouring into the basement. Upon further investigation, we found several other handyman practices in this home. The corrections to these problems cost the owner over $7,000 before the sale of the home was closed.*

#16 Handyman plumbing

- **Cross connections:** Watch for any possible cross connections where water from the drain and waste system can be siphoned back into the supply system. (See pages 36 and 37 for more information.) Remember to report any cross connections found as a **safety hazard**.

#17 Dangerous cross connection

*Photo #17 shows a **dangerous cross connection**. By the way, this is the same house belonging to the handyman mentioned above and was part of his $7000 plumbing fiasco. Here, he's got small black hoses from the water softening system sticking directly into the floor drain. Water from the drain can be siphoned into his water supply system. (See the paint can? His new P-trap was leaking.)*

Inspecting Drainage

The question home inspectors must answer for the purpose of the inspection of drainage is whether water drainage at fixtures and drains is **functional**. Does all the water drain from a fixture? Does it drain fast enough? The inspector should determine if drainage is satisfactory based on opinion and common sense. Drainage is adequate or satisfactory if the water drains fast enough and completely. If a sink takes forever to drain or if water doesn't drain completely, then drainage is poor.

The home inspector can test water drainage at the same time as testing water flow. This can be done in these ways:

1. As each faucet is run, just watch as water drains from the fixture.

2. Plug the fixture as water is running to fill the fixture, then pull the plug to watch the water drain.

There are different kinds of pumps that the home inspector may find in the basement that are considered to be part of the drainage system:

Personal Note

"Yes, I've ruined traps under sinks too. I can remember checking out a P-trap that was so rusted out that I put my hand right through it.

"It can be embarrassing. It doesn't matter if the trap was in incredibly poor shape to begin with. You end up having to pay for a new one to be installed."

Roy Newcomer

- **Sump pump:** When we speak of a sump pump, we're talking about a **fresh water pump** that pumps rainwater coming from the perimeter drainage system away from the house. The sump is a pit located below the basement slab. It can have concrete or earth walls and floor. Plastic liners may also be used. The electric sump pump is located in or above the pit to pump water from the sump. Water is pumped away from the house to a city storm sewer, onto the surface of the yard, or into a drywell. A **drywell** is a buried gravel pit that accumulates water and allows it to seep into the ground slowly. (Perimeter drain tile systems are presented in another of our guides — *A Practical Guide to Inspecting Structure*.)

The home inspector should test the pump for operation. It can be a **pedestal-style pump** which has the motor mounted on a shaft sitting above the water level. A lever will stick out of the crock. To test the operation of the pump, pull up on this lever. The sump pump may be the **submersible** type, sitting below the water level in the sump. To test this type of sump pump, use a wooden stick to pull up on the pressure switch or float in the crock. Another way to test each kind of sump pump is to run water into the crock with a hose (not required).

The sump pump motor should run quietly and should discharge water. It shouldn't run all the time. The pump should have its own dedicated circuit so it continues working even if some other equipment malfunctions. The crock should be covered and kept free of silt buildup and debris.

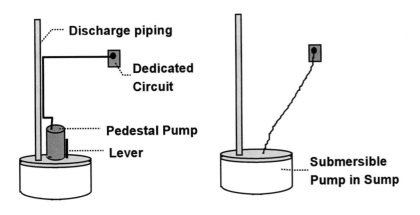

Personal Note

"One of my inspectors was inspecting an inner city property. While checking drainage at the kitchen sink, he found it to be extremely good. Water went out as fast as it flowed. The owner stood by and watched as the water ran.

"Finally, the inspector turned the water off and then bent down to inspect the trap and pipes under the sink. There weren't any drainage pipes at all! Instead, he found a large pail filled to the top. The pail had to be emptied elsewhere every time the sink was used.

"The owner shook her head and said, 'I wondered if you were going to let that overflow.' As if the inspector should have known."

Roy Newcomer

---

*Photo #18 shows a **pedestal-style fresh water sump pump**. You can see the pump lever to the left of the motor. This sump crock should be covered to prevent anyone from falling in.*

Guide Note

Information on foundation drain tile systems is presented in another of our guides — A Practical Guide to Inspecting Structure.

Personal Note

"One of my inspectors flushed a basement toilet and realized that the homeowner had the toilet draining to a fresh water sump pump instead of the septic system. The inspector later discovered that the garage floor drain also drained in that sump pump. "Now, you know that a fresh water sump pump is pumping water out into the back yard. In this case, the inspector took a look out there. He found waste from the toilet and oil the owner had poured down the garage drain on the surface of the lawn. An extremely unhealthy situation."

Roy Newcomer

#18 Pedestal-style fresh water sump pump

- **Sanitary pump:** The sanitary pump is like a sump pump in that it is a pit with a pump. However, water flowing into the sanitary pump is **gray water** or drainage from the clothes washer or laundry tub, sometimes from the floor drain. When the sewer or septic line from the house is above the floor level, the sanitary pump is needed to pump this waste water up to the sewer line level. A sanitary pump may discharge to an outside drywell instead of the sewer or septic system. This is also acceptable. However, pumping this gray water out onto the surface of the yard is not an accepted practice. Organic waste can rot and cause odors.

The sanitary pump crock should be covered and **sealed**. Some communities require the crock to be **vented**. There should be a **check valve** or backflow preventer valve in the discharge pipe itself. This check valve prevents waste from the sewer line from backing up into the crock. The valve is usually blue or green and has an arrow on it that points up, showing the proper installation of the valve. The home inspector should determine if the check valve is present.

Test the operation of the sanitary pump, if possible. You should be able to run water into the laundry tubs that will drain into the crock. Listen for the pump to kick on.

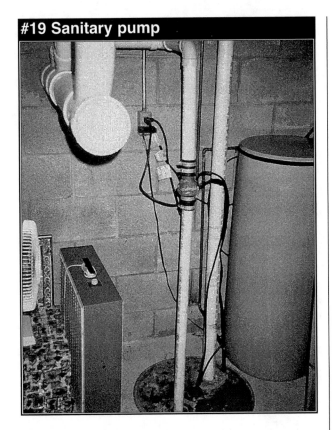
#19 Sanitary pump

*Photo #19 shows a **sanitary pump**. Notice the check valve in the discharge line.*

- **Solid waste pump:** When there is a toilet in the basement and the sewer line is above the floor level, a pump is required to discharge waste up to the sewer line. The solid waste pump is also submerged in a tank with a sealed top. A vent pipe is typically connected to the top of the tank. Some communities don't allow the use of the solid waste pump. The home inspector should be aware of local codes regarding their use.

Reporting Your Findings

As you're inspecting the DWV system, talk to your customer and let them know what you're doing. Be sure to explain just what you're inspecting and what your findings are. Take the time to answer questions. Remember that customers may not understand what they see in the mass of piping and are counting on you to make sense of it.

Be sure to review the inspection report with your customer after the inspection. Even though you've been careful to communicate during the inspection, often times the customer will forget some or all of what you've said. Go through the report page by page, pointing out where you've noted certain

Definitions

A <u>sump</u> is a pit located in the basement floor containing an electric pump. The pit collects fresh water from the footing drains and the <u>sump pump</u> pumps it away from the house.

A <u>sanitary pump</u> collects gray water from fixtures and pumps it up into the sewer line or <u>drywell</u>, which is a buried gravel pit. The sanitary pump has a <u>check valve</u>, which is a backflow preventer valve.

A <u>solid waste pump</u> collects waste from a basement toilet and pumps it up into the sewer line.

findings. This is especially important with technical systems such as plumbing, where the customer may not understand thoroughly. The review gives you another chance to test their understanding. Spend time pointing out where you've included major repairs, safety hazards, and items requiring replacement in the near future.

We've covered quite a bit of information about the DWV inspection (page 40 to this page) including waste disposal, venting, inspecting DWV lines, and drainage issues. These items should be reported on various pages within your inspection report. Again, it depends on the layout of the report you're using.

- **Kitchen:** After you test for **functional drainage** at the kitchen sink, rate the drainage as adequate or poor and record that fact in the inspection report. Note the condition of DWV piping under the sink and report defects such as leaking or corrosion. Watch for any illegal S-traps or mechanical vents (which may not be permitted in your area). This should be recorded. When reporting these defects, you can write your suggestions about remedying the situation. For example, if S-traps are present, you can suggest replacing them with P-traps and vents. Report the odor of sewer gas in this location.

- **Bathroom:** Report functional drainage in the bathroom on a bathroom page in the report. Again, rate it as adequate or poor. Report leaking, illegal traps, and so on as mentioned above.

- **DWV piping:** On the plumbing page of your report, identify the DWV piping materials used in the home. Note if you've found lead pipe used in the DWV line somewhere. Of course, lead DWV pipes don't pose the same hazard that lead supply pipes do, so don't issue a health hazard. Record the defects found such as those listed here and suggest a plumber should be called in for repair or replacement if your findings warrant it.

  — Corroded, rusting, damaged, and leaking pipes
  — Improperly supported or insulated pipes
  — Handyman or temporary fixes
  — Unsealed cleanout
  — Illegal traps or venting

NOTE: If you find DWV venting exhausting into the attic, leaks around the plumbing vent in the roof, or holes in vent pipes in the attic, report the conditions on the attic page of your inspection report. Sometimes, customers miss these details if they're reported on a general plumbing page.

- **Sumps and sanitary pumps:** For sumps, note whether or not one is present, whether or not it's been tested, and whether or not it's operating. For sanitary pumps, note if the crock is sealed, whether a check valve is present, whether or not you've tested the pump, and whether or not it's operating. Note any other defects you've found.

- **Safety hazards:** Again, always note safety hazards on the page of the report dealing with their location and again on a summary page at the back of the report. Record the following as safety hazards with DWV piping:

  — Cross connections between DWV and supply piping
  — Holes in vent pipes
  — Improperly installed vent pipes

- **Major repair and replacement:** You can classify **serious problems** with the DWV piping system indicating immediate and expensive replacement as major. But be careful not to highlight every little finding as a major repair or replacement. Sump pumps and sanitary pumps that are not operating and need replacement should be put in this category.

SUMP PUMP LIFETIME

Since sumps have a relatively short lifetime, it's a good idea to list them as items needing replacement within the next 5 years. Just for your protection.

Personal Note

"Here's another example of improper waste discharge. One of my inspectors flushed a toilet in a remodeled basement and thought it sounded weird. He put toilet paper in it and flushed it again. Then he looked into the fresh water sump pump. Sure enough, he found the toilet paper in there.
"Watch out when you find a new bathroom in a remodeled basement. Homeowners sometimes foolishly take these shortcuts for waste discharge."
Roy Newcomer

WORKSHEET

Test yourself on the following questions.
Answers appear on page 60.

1. DWV pipes are never truly horizontal because:

   A. They're too heavy to hold up.
   B. They only run vertically in the house.
   C. They require a slope for drainage purposes.
   D. They're not manufactured that way.

2. Which of the following is the home inspector <u>not</u> required to inspect?

   A. Septic system
   B. Sump and sanitary pumps
   C. Sink traps
   D. Drainage at plumbing fixtures

3. What combination of DWV piping is often found in homes built in the 1950's?

   A. Copper and chrome-plated brass
   B. Galvanized steel, lead, and PVC
   C. Cast iron, ABS, and PVC
   D. Cast iron, galvanized steel, and chrome-plated brass

4. How is horizontal cast iron piping vulnerable?

   A. The top of the pipe rusts from water sitting in the pipe.
   B. Sewer gas can eat through the top of the pipe.
   C. It can easily split along the top.
   D. Joints can easily break.

5. What does the term *wet vent* mean?

   A. A vent that's leaking
   B. A vent that carries both air and water vapor
   C. A vent that carries air, water vapor, and waste matter
   D. A vent that exhausts into the attic but not through the roof

6. Which of the following traps may be illegal according to local plumbing codes?

   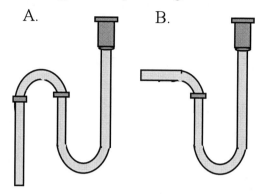

7. A drain that gurgles or belches as it drains water may be a symptom of:

   A. The presence of a mechanical vent
   B. A leak in the trap
   C. A missing vent
   D. The presence of a drum trap

8. What finding is considered a safety hazard?

   A. A corroded P-trap
   B. The use of copper DWV piping
   C. Missing insulation on pipes
   D. Holes in a vent pipe

9. What should the home inspector watch out for if a plastic S-trap is found under a sink?

   A. Cross connections
   B. Handyman plumbing
   C. A private septic system
   D. Poor functional drainage

10. If the sewer line is above the level of the basement floor, what type of pump should be used to discharge gray water from the clothes washer and laundry tubs?

    A. Sump pump
    B. Sanitary pump
    C. Solid waste pump

Chapter Five

HOT WATER SYSTEM

The hot water system in most homes is the common **water heater**, although some have systems associated with the home heating equipment such as a boiler or heat pump. During the inspection of the water heater, the home inspector identifies the **brand** of water heater, its **type** (gas, oil, or electric), and its **capacity** in gallons. The inspector also estimates its **approximate age**. The inspection includes the following:

- The condition of the water heater tank
- Proper piping, venting, and location
- Operating and safety controls
- Operation of the water heater

Water Heaters

A water heater is a tank, usually steel-lined on the inside with glass, porcelain, or cement and wrapped on the outside with insulation and an enameled metal jacket. Water heaters vary in size. Some standards suggest an 80 gallon tank for a family of 4 with a clothes washer and dishwasher, although a 30 to 40 gallon gas or oil water heater or a 40 to 50 gallon electric one are common. An electric water heater has a slower **recovery rate** than oil and gas. A gas water heater has an hourly recovery rate about the same as the tank volume. Oil water heaters recover even faster.

Whether fueled by gas, oil, or electricity, water heaters have the following components in common:

- **Magnesium anode rod(s):** Most tanks have 1 or 2 magnesium anode rods inside to protect exposed steel by giving up ions. (See pages 23 and 24 on electrolytic corrosion.) In general, a tank with a 5-year guarantee has 1 rod; a tank with a 10-year guarantee has 2 rods.

A water heater tank can become contaminated with **Desulfovibrio bacteria**, which thrives in hot water in the presence of magnesium. The bacteria cause the hot water supply (not the cold) to smell like rotten eggs. If this smell is noticed, the water can be tested for contamination. The

Guide Note

Pages 59 to 71 present the study and inspection of the hot water system.

HEATER COMPONENTS
• Anode rods
• Cold water turn-off valve
• Inlet, dip tube, and outlet
• Relief valve and extension
• Operating controls and thermostat
• Drain valve

solution to the problem is to chlorinate the tank and replace the anode with an aluminum rod.

- **Cold water turn-off valve:** The incoming water supply pipe to the water heater comes from a water softener or conditioning system or from the main line. There should be a cold water turn-off valve on the supply side near the water heater so the water can be turned off in case of repair or replacement. (Do not operate this valve during the inspection.)

- **Inlet, dip tube, and outlet:** The **inlet** on the top of the tank sends cold water to the bottom of the tank through a **dip tube**. That keeps the cold water from cooling down the hot water which rises normally to the top of the tank. Hot water exits the tank from the **outlet** at the top. Sometimes, you'll find that the inlet and outlet pipes are reversed. Then cold water enters right into the top of the tank because it isn't going through a dip tube. Someone who takes a shower is going to start off with a blast of cold water. However, sometimes plumbers reverse the inlet and outlet pipes for convenience but correct the situation by also moving the dip tube to the outlet.

- **Relief valve and extension:** All water heaters must be protected with a relief valve, also called a temperature-pressure relief valve (TPR). This valve lets water escape if the temperature or pressure is too high in the tank. It should be mounted on the top of the tank or on the side within 6" from the top. You may find relief valves mounted on the hot water pipe above the tank, but this practice is no longer acceptable and is not considered safe. A leaking relief valve must be replaced. (Do not trip this valve during the inspection.)

The relief valve should have a **metal extension** or discharge tube that extends down the side of the tank to within 6" of the floor so hot water and steam won't spray anyone nearby. The extension may not be threaded at the bottom or have a turn-off valve. This is to prevent anyone from capping or plugging the extension.

- **Operating controls and thermostat:** Water heaters are designed to output water at 140° and can be set to individual preferences with the thermostat. Dishwashers

Worksheet Answers (page 58)

1.	*C*
2.	*A*
3.	*D*
4.	*B*
5.	*C*
6.	*A*
7.	*C*
8.	*D*
9.	*B*
10.	*B*

may require a temperature of 140°, but a range from about 115° to 120° is safer and more efficient. Gas water heaters have a dial or knob type thermostat at the control unit, which also has gas control devices. The thermostats on electric heaters are adjusted by using a screwdriver.

- **Drain valve:** Every water heater should have a drain valve at the base of the tank. A few gallons of water should be drained out several times a year to get rid of any sludge or sediment buildup in the tank.

The following rules apply to the installation of water heaters in the home:

- **Location:** If the water heater is located **outside**, it should sit on a concrete pad with its base at least 3" above grade. If located in the **garage**, a gas or oil fueled water heater should sit at least 18" above the garage floor to prevent ignition of gasoline vapors in the garage. If located inside the **living area** of the home on a wooden floor, the water heater should be installed with a drip pan under it. Gas and oil fueled water heaters should not be installed in closets, bedrooms, or bathrooms. These types of fossil fueled heaters need to have an adequate and continuous air supply for proper combustion.

- **Exhaust:** Fossil fueled water heaters must be exhausted to the outdoors, usually through the chimney, although some are designed to vent directly through the wall to the outside. Carbon monoxide is a life-threatening byproduct of combustion and must be safely exhausted.

Gas and oil water heaters must have a **flue pipe** from the tank to the chimney, requiring an upward slope of 1/4" for every foot of pipe. If the flue pipe is a single-wall pipe, it should be installed at least 6" away from combustibles. . The double walled flue should vent a minimum of 1' above the roof (2' for single wall) or higher depending on the slope of the roof. A **draft hood** is required at the top of the heater to prevent backdrafts from the chimney sending carbon monoxide back into the home.

In general, water heaters last only about 8 to 12 years, possibly longer depending on their size, maintenance, and use. Faulty valves and controls can be repaired, but any water heater that leaks must be replaced.

> **NOTE**
>
> The home inspector does not have to inspect a water softener or conditioning system. Just confirm that it's hooked up and there's salt in it.

> **ANOTHER NOTE**
>
> Check the BTU reading on the relief valve tag. It must exceed the BTU reading on the water heater data plate.

Gas Water Heaters

The following procedures should be followed when inspecting a gas water heater:

1. **Examine the manufacturer's plate** to find the brand name and capacity of the water heater. Look at the serial number, which typically gives an indication of the year of manufacture. If the serial number doesn't alert you to the age of the heater, ask the owner when the water heater was purchased. You may not always get a truthful answer. Record the age in your inspection report as the *approximate* age.

2. **Notice installation** and report any infractions of the rules. Gas water heaters should not be closed off in closets or located in bedrooms or bathrooms. They must have good air circulation and not be in areas that can endanger occupants of the home.

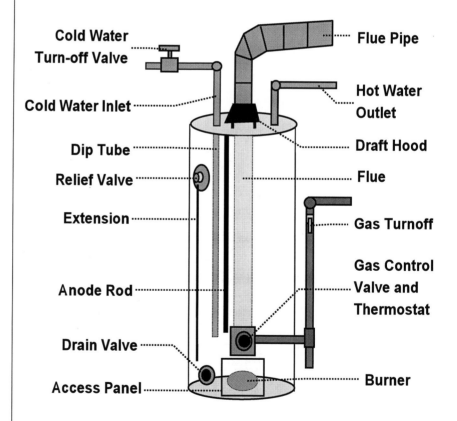

Cold Water Turn-off Valve

Cold Water Inlet

Dip Tube

Relief Valve

Extension

Anode Rod

Drain Valve

Access Panel

Flue Pipe

Hot Water Outlet

Draft Hood

Flue

Gas Turnoff

Gas Control Valve and Thermostat

Burner

3. **Check inlet and outlet pipes**, feeling them to confirm that cold water supply pipes are connected to the marked inlet and hot water distribution pipes are connected to the outlet. If you find the pipes reversed, make a note of it in the inspection report. Explain to the customer what the implications of the reversal are (pulling cold water out of the top of the tank when turning the hot water on). Also explain that a plumber may have moved the dip tube to the outlet to compensate for the situation, although you have no way of checking that.

#20 Plastic water supply pipes

*Photo #20 shows **plastic water supply pipes** to and from the water heater. It's all right to use CPVC plastic here, even for the hot water supply, but in this case the plastic pipes are too close to the flue pipe coming from the water heater. The flue pipe gets too hot. During the inspection, you should check the cold water supply pipes for a turn-off valve so you can point it out to the customer. In this photo, the valve is missing. We noted that in our inspection report, along with a caution about having the CPVC too close to the flue.*

4. **Turn down the thermostat** located at the bottom of the tank on the control unit in order to turn down the flame. This is for your own protection when you remove the burner plates. CAUTION: Make a mental note of the thermostat setting before you turn it down. You'll have to return it to this same setting after the inspection.

5. **Remove both burner plates** (access panel and interior burner plate) located below the control unit.

6. **Inspect for rust, corrosion, signs of leaking, and sediment buildup.** Be sure to check inside the heating chamber and watch for rusting, corrosion, and any

For Beginning Inspectors

If you have a gas water heater, take this book with you and go inspect the water heater as laid out on these pages. Be sure to follow each step carefully for your own protection.

The home inspector checks and reports on adequate hot water flow at the faucets. Although this is directly related to the hot water system, we'll be discussing this on pages 73 to 74 of this guide.

*   Photo #21 shows* **heating chamber problems**. *This tank is corroding both inside the heating chamber and at the opening around the chamber. We felt underneath this tank and discovered that the bottom of the tank was corroded and leaking. We advised the customer that the water heater had to be replaced.*

sediment buildup that can impede the efficient operation of the flame. Then inspect the bottom of the tank and the area around the access panel for rust, corrosion, and leaking. Feel along the bottom of the tank for leaks too. If you find water dripping from the tank, try to determine if the tank is indeed leaking or whether you've got leaks from the drain valve or some other connection. If the tank is leaking, it's shot and must be replaced.

#21 Heating chamber problems

7. **Turn up the thermostat** so the flame kicks on. You may also want to run some hot water at this time to be sure the flame ignites. **Evaluate the flame**, which should burn mostly **blue** or blue with some orange. Flames that burn **yellow** and **flame rollout** are serious problems. Flame rollout is when the flame burns out from behind the burner plate to the outside of the tank. If you find these conditions, recommend that a specialist be called to fix the situation. Take another look at **Photo #21.** The soot outside the tank above the heating chamber indicates a flame rollout problem.

8. **Listen to the tank** as the burner is firing. A **thumping or rumbling noise** is an indication of sediment buildup inside the tank. You can suggest to the customer that sediment and sludge can be drained off through the drain valve, which you should point out. Recommend that a few gallons of water be drained from the tank several times a year to help prevent this buildup.

9. **Replace the burner plates and reset the thermostat.** CAUTION: Be sure to reset the thermostat to the **same temperature** you found it. Forgetting to do this is high liability for home inspectors. You don't want to be the cause of scalding any occupants of the home by leaving the water temperature too high.

10. **Smell for gas leaks** around the gas control valve and at unions and connections to gas lines near the water heater. There are electronic testers available on the market that will detect the presence of natural gas in the air. However, these testers often detect minute quantities of natural gas that gas company inspectors would consider a safe level. The home inspector should tell the customer about any faint gas odor and, if the odor is *extremely* strong, should have people leave the house immediately and call the gas company to come in and fix the problem. Gas leaks should be reported as a **safety hazard**.

11. **Inspect the relief valve and extension.** The relief valve should be mounted directly into the tank, either on the top of the tank or at the side within 6" of the top. The valve should be free of corrosion and should not leak. By the way, *don't trip this valve during the inspection.* And report a missing or improperly installed relief valve. Missing relief valves should be reported as a **safety hazard**. Leaking or corroded valves should be replaced.

Check that extensions are present and properly installed. Most communities require the extension to be copper or galvanized, although some will allow CVPC. The extension piping should not be downsized as it goes down the tank. Check that the extension ends about 6" from the floor and is not threaded or capped at the end.

Photo #22 shows a **relief valve** connected into the tank at the top. It's located just to the right of the cold water supply pipe. Here, plastic piping is used for the extension piping. Remember that some local codes don't allow CVPC piping for the extension. If you're area doesn't allow it, this situation should be reported and a suggestion made to replace the extension piping with copper or galvanized.

#22 Relief valve

Photo #23 shows an **unacceptable relief valve**. Here, it's mounted on a pipe on top of the tank, which is no longer considered a safe condition, not into the tank itself as it should be. You can see the valve at the left of the flue. There's something wrong with the extension here. Do you know what it is? First of all, the extension, instead of going to the floor and ultimately to the floor drain, goes into the laundry tub below the fill level in the tub. That creates the possibility of a dangerous **cross connection**. Water from the tub can get sucked into the water heater, contaminating the water supply. This was reported as a **safety hazard**. Secondly, not only is the extension at the bottom **threaded**, there's an elbow connected to it that would shoot hot water and steam sideways. This was also reported as a **safety hazard**.

#23 Unacceptable relief valve

12. Finally, inspect the venting. First, check for the presence of the **draft hood**. Light a match and hold it next to the draft hood. Move it around the hood, watching to see which way the flame leans. If the flame leans in toward the hood, that's good. But if the flame leans outward from the hood, there may be a downdrafting condition, indicating that carbon monoxide is being released from the hood. This should be reported as a **safety hazard**. An inspection mirror can be used to test for downdrafting too. Hold the mirror close to the hood. A fogged up mirror indicates downdrafting. An electronic tester may also be used to check for the presence of carbon monoxide.

Inspect the **flue pipe** for proper installation and to note any rusting or corrosion on the pipe. Any holes in the flue pipe should be reported as a **safety hazard**.

#24 Leak above the water heater

*Photo #24 shows a **leak above the water heater** that is causing damage. Notice the leaking drum trap above the heater. This had been dripping onto the heater for some time. Notice the rusting and corrosion at the top of the tank and around the draft hood. We checked the top of the flue pipe in this area to be sure that the leak hadn't caused the pipe to develop any holes from rusting and corrosion, which would be a safety hazard. There's one other thing to notice in this photo. The relief valve is in the cold water supply pipe at the right without any extension at all, a safety hazard. It should be connected directly into the tank. We recommended that a plumber come in to repair the leak and to install a proper relief valve and add an extension.*

"An instructor here at the American Home Inspectors Training Institute mentioned to me that he once came across an electric water heater that was 47 years old. It was an old 120 gallon Westinghouse without any significant problems. It was still going fine."

Roy Newcomer

Electric Water Heaters

Electric water heaters typically have 2 **heating elements**, one at the top of the tank and one at the bottom. They generally don't come on at the same time. The top unit comes on when the tank is cold. When the top 1/4th of the tank is heated, the lower element will come on.

The heating elements are connected as a series. That is, if the top element is burned out, then the bottom element won't function. However, if the bottom element is burned out, the top one will still function. If there's no hot water at all, that means the top element is blown. If there's limited hot water, it could be that the bottom element is shot.

The inspection of the electric water heater is similar in some ways to that of the gas heater, but different in others. Follow these procedures during the inspection:

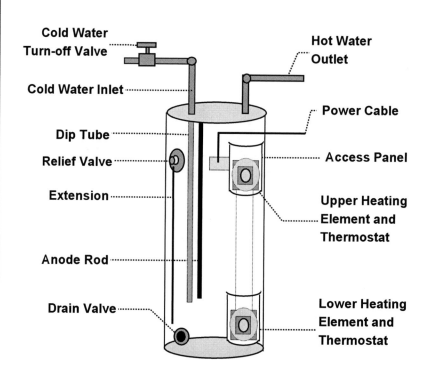

1. **Examine the manufacturer's plate** for information about the tank.

2. **Notice installation.** Does an electric heater on a wooden floor in the living area have a drip pan?

3. **Check the inlet and outlet pipes** and report reversed pipes. Is there a cold water supply turn-off valve?

4. **Inspect the tank for rust, corrosion, and leaking.**

5. **Remove upper and lower access panels.**

6. **Inspect for any wiring problems or rust and corrosion** on the heating element. Wiring problems in the heating element may include conditions such as overheating or corroded wires, but this generally is uncommon.

7. **Feel the tank at the heating elements.** You don't have to adjust the thermostat to test whether the heating elements are working. Just put your thumb against the tank at each heating element to see if the tank is hot. If the tank is cold at the top, the upper element is burned out and the bottom element would not be working. If the tank is cold only at the bottom, the lower heating element may be burned out.

8. **Turn on a hot water faucet** and let it run for a while. This is to activate the heating elements.

9. **Listen to the tank.** When the heating elements come on, listen for a hissing or gurgling sound which would indicate the presence of sediment buildup in the tank.

10. **Inspect the relief valve and extension.** The same conditions should be met as we've discussed earlier on page 65 relating to gas water heaters.

Other Hot Water Systems

Some homes may have other types of hot water systems. One example is the **tankless coil** associated with the heating boiler or a stand alone model. The tankless coil may be located inside the boiler or adjacent to it. One disadvantage with tankless coils is that the boiler must be kept running in the summer. There may be too little hot water supplied with these systems or the coils can become clogged with silt and salts from the water. Water inside the boiler is heated to 190°, which is very dangerous at the faucets, so a mixing valve should be present to mix cold water with the hot water and thereby lower the temperature to an acceptable range.

HEATING ELEMENTS

If the upper heating element is burned out, the lower won't function. There won't be any hot water available at the faucets. If the lower element is burned out, the upper will still function, but the hot water supply will be limited.

Personal Note

"Here's an interesting story. A home had been closed up for the winter and the water heater emptied. The homeowner came back to dewinterize the home. He fired up the electric water heater before it had a chance to completely refill with water — it still had lots of air in it. When he turned it on, the top element exploded and blew right out of the tank."

Roy Newcomer

Reporting Your Findings

Turn to the appropriate page of the inspection report to record your findings on the water heater.

Water heater information: You should record the brand name of the water heater from the manufacturer's plate, the serial number, and its capacity. Identify the fuel source too (gas or electric). Next, write the approximate age of the water heater as best as you can figure out (usually the first 2 numbers in the serial number). We suggest that you caution your customer that the water heater may have to be replaced within 5 years if it's older than 5 years at the present time.

Water heater condition: Note defects found on the water heater such as a rusted or leaking tank, burned out heating element, reversed water lines, sediment buildup, and improper operation or venting.

Safety hazards: Be sure to note safety hazards such as vent problems (downdrafting at draft hood and missing or corroded flue pipes), gas leaks, a missing relief valve, and a missing or improper extension.

WORKSHEET

Test yourself on the following questions.
Answers appear on page 72.

1. What is the home inspector <u>not</u> required to do while inspecting a hot water system, according to the standards of practice?

   A. Inspect an electric water heater.
   B. Inspect the relief valve.
   C. Operate the relief valve.
   D. Describe the water heating equipment.

2. What is the purpose of an anode rod in a water heater?

   A. It sends cold water to the bottom of the tank.
   B. It gives up ions before the steel tank does.
   C. It turns off water so the tank can be repaired.
   D. It keeps the water temperature at 140° in the tank.

3. Identify the following components of a gas water heater.

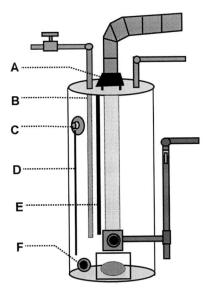

4. A gas water heater may be vented to the attic or the garage.

   A. True
   B. False

5. Which statement is true about the installation of a gas water heater?

   A. It may be installed in a closet.
   B. If located in a garage, it must be installed 3" above the floor.
   C. If located outside, it may be installed on the ground.
   D. If located in the living area of a home, it must have a drip pan.

6. What condition with a gas water heater should be reported as a safety hazard?

   A. Holes in the flue pipe
   B. A leaking tank
   C. Rusting in the heating chamber
   D. Plastic water supply pipes

7. What does hissing and gurgling in an electric water heater indicate?

   A. The upper element is burned out.
   B. The lower element is burned out.
   C. There's sediment in the tank.
   D. The heating element wires are overheated.

8. What happens to an electric water heater when the upper heating element burns out?

   A. The lower heating element won't function.
   B. The lower heating element has to work extra hard to heat the tank.
   C. There's only limited hot water available at the faucets.
   D. The water heater explodes.

9. Which component is present on a gas water heater but not on an electric one?

   A. A power cable
   B. A dip tube
   C. A draft hood

Guide Note

Pages 72 to 80 present the study and inspection of fixtures and faucets.

Chapter Six

FIXTURES AND FAUCETS

The home inspector inspects each fixture for its **condition** and runs each faucet in the home, testing for **drips and leaks**. Functional flow and drainage and traps are also checked, which were discussed earlier in this guide.

Inspecting Faucets

The traditional **washer faucet** has a washer attached to a threaded stem which is turned down and forms a seal against a mechanical seat. This type of faucet supplies either cold or hot water. **Washerless faucets** are commonly single lever faucets supplying a mix of cold and hot water. They have a cartridge, valve, or ball, with or without an O-ring seal, which move up and down for volume and left and right for hot and cold.

There are more sophisticated **shower faucets** that will maintain a constant temperature, even if there are pressure changes in the system. If cold water flow drops, a pressure-sensitive mixing value in the faucet will adjust hot water flow to maintain the same temperature.

The home inspector should **operate all faucets** in the home, including exterior hose bibs, during the plumbing inspection. The inspector should watch for the following conditions:

• **Leaks and drips:** Notice any leaking or dripping at each faucet before you turn it on. Sometimes, stains in the fixture or around the faucet stem can indicate a leak problem. Watch for leaking when the faucet is running.

— **Leaks and drips out of the faucet:** In the washer-type faucet, this usually indicates a deteriorated washer, which is inexpensive to replace. But it could be a damaged seat. For other faucets, it could be a deteriorated O-ring.

— **Leaks from the faucet handle or stem:** In the washer-type faucet, this condition can be caused by a worn or bent stem, deteriorated stem packing, or a loose stem packing nut. In other faucets, this indicates faulty valves or cartridges, which will have to be replaced.

— **Leaking behind or under the faucet:** If the faucet is not tightly secured to the fixture, countertop, or wall, splashed water can run under it, causing staining and damage to walls and floors.

Leaky faucets should be pointed out to the customer and reported in your inspection report.

- **Noisy faucets:** Faucets should run smoothly and quietly. **Whistles** while the water is running indicate a faulty interior design in the faucet and is hard to fix. **Chatter and banging** while water is flowing is probably a loose washer. **Water hammer**, which occurs in the pipes just as the faucet is turned off, is not a faucet problem (see page 31). Noisy faucets should be pointed out to the customer during the inspection.

- **Damaged or corroded faucets:** Check each faucet for any damage or corrosion. Ceramic faucet handles can become cracked, and the sharp edges can cut hands. Be sure to report any you find. Corroded faucets may be so damaged that the home inspector should recommend replacement.

- **Loose faucets:** Check faucets to see if they are properly tight at fixtures, countertops, and walls in order to prevent leaking into the wall or floor structure.

- **Cross connections:** Always check out the position of the faucet and be sure that it's located above the fill level of the sink or tub to avoid a possible cross connection (see pages 36 and 37). Be sure to record potential cross connections in your inspection report and report each such case as a **safety hazard**.

- **Poor functional flow and lack of hot water:** See the discussion on pages 34 and 35 for testing the plumbing system at the faucets for functional flow. When running faucets, use your own judgment on the adequacy of water flow. It's possible that the flow of water at only a single faucet in the house is not good. Many new faucets have built-in pressure reducers allowing only a 2 to 3 gpm flow. But there may be an internal problem with the faucet — the spout could be clogged with corrosion or mineral deposits. Aerators on faucets, which add air to the water flow, can become clogged with salt deposits. Suggest to the customer that aerators be removed and cleaned.

INSPECTING FAUCETS

- Leaking and dripping
- Noisy faucets
- Damaged or corroded
- Loose faucets
- Cross connections
- Poor functional flow

Guide Note

Outdoor faucets or hose bibs must also be checked during the plumbing inspection, if not in a winterized condition. This was discussed earlier in the guide. See page 32.

Be sure to check each hot water faucet or combined faucet for the **presence of hot water**. Pay attention to the temperature of the water. If the water is too hot, you may want to suggest to the customer that the thermostat on the water heater should be set at a lower temperature to prevent accidental scalding. A safe setting is between 115° and 120°. Report the absence of hot water at any hot water faucet.

Inspecting Sinks, Tubs, and Showers

Sinks, basins, and bathtubs can be made of enameled steel or cast iron, stone, or plastic. Those made of fired clay include porcelain, ceramic, and china. A bidet, which is really a type of sink, is usually made of china. Showers may be tiled enclosures with tile floors over metal shower pans or made of up to 3 pieces of molded plastic or fiberglass. Each fixture should be inspected during the plumbing inspection. The home inspector should watch for the following conditions:

- **Broken, damaged, or rusted fixtures:** Any fixture that is actually broken or cracked completely through should be reported to the customer. Ceramic fixtures can become nicked or chipped. Porcelain can get a network of hairline cracks called crazing over time. Enameled finishes can get nicks, spalled spots, or damaged finishes from harsh scouring cleaners. Plastics can crack, distort from poor installation or become abraded. Any fixture may become stained from rust or copper. Point out these conditions to your customer as you inspect the fixtures and note them in your inspection report.

Always look under sinks during the plumbing inspection, even if enclosed in cabinets. Look at the bottom of enameled steel sinks for any signs of rusting. They can rust out at the welded steel overflow and will eventually leak.

- **Missing overflow:** Kitchen sinks and laundry tubs don't generally have overflows, but bathroom basins and tubs typically have them. Point out their absence to the customer with a reminder that one would have to remain in the bathroom to prevent overflow.

- **Loose mountings:** Check that sinks and basins are properly secured to the walls or to supporting columns or countertops. Loose fixtures can damage both the fixtures and the supply and drain piping. Tubs are supported by the house framework and may shift if the structure moves or settles. Report such conditions.

- **Poor functional drainage:** The home inspector should run water at each fixture and observe functional drainage (see page 52). Check the drain fittings at this time to see if they're operating properly. Plug the fixture and let it hold water for a time to see if the plug holds water.

- **Leaks at and around fixtures:** Always check under sinks for any evidence of leaking, either from the fixture or faucets, from the supply piping, at the drain fitting, at the traps, and in the DWV piping. One problem to watch out for is the spray hose connection at the kitchen sink, which often leaks.

  Carefully check the floor around the bathtub and be sure to report evidence of **wood rot** from leaking. Tubs can leak at the tub overflow and drain fitting, which would be noticed only from a stained ceiling underneath or in the basement under a first-floor tub. Bathtubs with tile surrounds are vulnerable to leaking at the tub-tile intersection where caulking may be inadequate. Watch for leaking in the tile surround itself and note whether grouting is in good condition. Take another look at **Photos #14 and #24** for examples of evidence of leaks under bathtubs.

  Check for leaking at showers too, especially those with tiled walls that may have the same caulking and grouting problems as tubs do.

- **Metal shower pans** are a special problem. Although modern plastic and fiberglass molded shower pans are fairly good, the old metal pans are notorious for leaking. These shower pans were made of lead or tin and were covered with a finishing layer of ceramic tile. In fact, you should suspect a metal shower pan if the floor of the shower is finished with ceramic tile. The shower begins leaking through the tile grout around the drain, water gets into the metal pan, and the pan rusts through. They usually fail completely in 10 to 15 years. Never miss reporting the **presence of a metal shower pan** and pointing out the potential problems to the customer.

Personal Note

"Never miss the presence of an old tin or lead shower pan. Even if you don't see leaking now, you can be assured it's going to happen. Warn customers that they just don't last very long and can cause problems."

Roy Newcomer

If you're not sure the shower pan is leaking, cover the shower drain and let water fill the bottom of the shower. Then uncover the drain and try to observe leaking, if possible, from underneath.

#25 Shower with a ceramic tile floor

*Photo #25 shows a **shower with a ceramic tile floor.** We suspected a metal shower pan and our suspicions were confirmed.*

#26 Result of leaking

*Photo #26 shows what we found and the **result of leaking** from under the shower pan. You can see rotted wood and delaminated plywood all around this area. Notice the corrosion at the shower drain and all the water stains below.*

- **Improper traps or venting:** When you look under sinks, inspect the traps as described in page 49 (see pages 43 and 44 for additional information).

NOTE: Bidets should be inspected too. Bidets, because water supply is below the fill level, should have a vacuum breaker preventing waste water from flowing into the supply line. China bidets can crack and leak. The small jets can become clogged. Run water into the bidet and observe functional flow and drainage.

Inspecting Toilets

Toilets are normally made of porcelain enamel. There are 4 common types. The old **washdown toilet** is identified by the large bulge in front of the bowl and a relatively small wetted area in the bowl. It's no longer commonly used and some local codes don't allow them at all. The **reverse trap** is a better 2-piece toilet (tank and bowl) with a larger wetted area. The **siphon jet** is an improved type of reverse trap toilet with a quieter flush. The 1-piece **siphon vortex** or silent flush toilet is very quiet and has almost the whole bowl area covered with water.

The home inspector should **flush all toilets** in the home during the inspection. If an unused toilet is found, such as one in the basement, note if the water supply is turned off. Toilets should not be left sitting without water in the bowl as this would let sewer gas escape into the house. Report any unused toilets and suggest that water or some other liquid be kept in the bowl at all times.

Watch for the following conditions while inspecting each toilet in the home:

- **Damaged, cracked, or leaking:** Inspect the toilet carefully for cracks in the bowl or tank. Report even those cracks on the interior of the bowl and tank lid that aren't the cause of any leaking. Report chips and nicks, crazing or hairline cracks, and any discoloration or serious staining.

  As to leaking, toilets can leak from the supply line, at the storage tank, at the connection between the tank and the bowl, at the bowl itself, and at the connection between the bottom of the toilet and the drain pipe. Look in each area for any evidence of leaking. Toilets that leak through a crack must be replaced.

Personal Note

"One of my inspectors was inspecting an old house with lead supply piping throughout. At the toilet, he was checking the bolts that hold the tank. He touched one to discover that the bolt was missing and the hole was self-healed with corrosion. "Of course, his touching it caused the corrosion to break through and the tank sprung a vigorous leak that couldn't be stopped. I can still hear him saying, 'All I did was touch it.'"

Roy Newcomer

AGE OF THE HOME

The date of manufacture is usually imprinted on the underside of the <u>toilet tank lid or tank back</u>. This gives a clue as to how old the house is, if it still has the original toilets.

Personal Note

"Nothing ever surprised me more than when I pulled up on a toilet and found myself standing there holding it in my arms. You don't have to nudge or pull on it very hard to tell if it's loose."

Roy Newcomer

Be sure to look for and report any evidence of **wood rot** around the base of the toilet. Get down on your hands and knees and push on the floor around the toilet to see if there are soft spots. It's not unusual to find rotted floor boards. The toilet could have experienced prior leaking and been repaired but left considerable damage to the subflooring. Be sure to report the presence of any wood rot.

- **Poor operation:** A toilet that won't flush at all should be reported as not operating. Often, the toilet has what is called a **weak flush** where waste is carried away in a sluggish manner or not all the waste is carried away before the end of the flush. This condition may have one of many causes. One cause is that there is too little water volume in the bowl due to a problem with the ball-cock mechanism in the tank or the holes around the bowl rim blocked with sediment. Weak flush may also be caused by downstream problems such as blockage of the toilet trap, vent stoppage, an improper slope of a waste pipe, or a clogged waste pipe.

Another operational problem is a **running tank** which continuously sends water into the bowl. This condition is caused within the tank and repairs should be made to the component of the flush mechanism causing the problem.

Mention both conditions — weak flush, running tank — to your customer and be sure to report them in your inspection report.

- **Loose toilet:** During the toilet inspection, check to see if the bowl is tightly secured to the floor. First, straddle the toilet bowl. Then, either hold the bowl between your knees and gently rock it back and forth or grab it with your arms and gently nudge or lift the bowl. But not too hard! The reasons a toilet can be loose can be improper fastening to the floor plate, rotten subflooring, or sunken joists underneath. A loose bowl can damage the wax sealing ring between the porcelain and the toilet bend, which will then cause leaking. Explain to the customer that the bolts on each side of the toilet bowl should be kept tightened so the seal doesn't break. Be sure to report any loose toilets that you find.

Putting It All Together

In this guide, we've presented the inspection of the plumbing system by components, not by the flow of how the plumbing inspection would really happen. Let's review how the inspection will be performed in proper order:

- **Exterior plumbing:** While inspecting the exterior of the house and the garage, you would operate outside faucets, climate permitting. Check for the proper location, extension, and flashing for the plumbing vent when you're on the roof. Notice the evidence of a private well and septic system and check for proper distances.

- **Interior living area plumbing:** Starting in the kitchen, run the water at the kitchen sink, checking the condition of the faucets, the sink, the sink traps and pipes under the sink. Note the presence of hot water, water flow, and drainage. In the bathroom(s), run water at more than one faucet simultaneously to determine if water pressure and flow are adequate. Check faucets and fixtures for damage, corrosion, the presence of hot water, and drainage. Always watch for any signs of leaking, especially around tubs, showers, and toilets. Test the toilet for operation. Report the presence of a metal shower pan. Keep an eye out for any signs of leaking or water stains.

- **Attic area:** While you are inspecting the attic, check that the plumbing vent is exhausted through the roof and is in good condition.

- **Basement:** Here's where you'll be able to get a good look at the underside of the upstairs plumbing. You will have run enough water through the drain and waste pipes earlier in the inspection to see any *active* leaks, but look for old leaks (stains) too. Now you'll inspect the water supply entrance, supply distribution piping, DWV piping, well equipment, sump and sanitary pumps, and water heater.

INSPECTION ORDER

- Exterior faucets, roof vent, location of well and septic

- Kitchen and bathroom faucets, fixtures, traps, piping, water pressure and drainage

- Living area signs of leaking

- Attic plumbing vent

- Basement supply entrance and piping, DWV piping, well equipment, sump and sanitary pumps, water heater

Reporting Your Findings

Don't bypass plumbing fixtures and faucets as too insignificant to report on. These are just the details you'll want to emphasize in your report. Now, let's get back to reporting on your inspection of the plumbing fixtures and faucets. New homeowners often call the home inspector when a faucet is leaking, thinking we're responsible. If you've reported the leaking faucet, you're not responsible. But if you've missed it, you may be. It's as simple as that.

• **Kitchen:** Report on faucet and sink conditions and functional flow and drainage and the presence of hot water.

• **Bathroom:** Report on faucet conditions in the bathroom, including functional flow and drainage and the presence of hot water. Be sure to record any evidence of leaking or water damage in bathroom ceilings and floors. Your report should have distinct lines for reporting on the toilet, tub, and shower conditions. Be sure to note if the toilet bowl is loose and whether the toilet operates. If you've found a metal shower pan, be sure to write a special note with a warning that if it's not leaking already, it will most likely do so in the future and will need replacement.

Chapter Seven

GAS PIPING

Some standards include the inspection of **fuel storage and distribution piping** in the plumbing inspection. For the most part, that means natural gas lines in the home and the storage of bottled gas.

The indoor piping for natural gas and bottled gas, also called LP (liquified propane or petroleum) gas, is usually **iron or steel**. Copper and brass are not allowed if gas contains >0.3 grains of hydrogen sulfide per 100 cu ft. (check with supplier). In some areas, flexible copper piping may be permitted for natural gas at connections between metal lines and appliances such as the kitchen stove. Plastic may or may not be allowed, depending on local codes. Galvanized steel is allowed in some areas. Rubber tubing is not allowed anywhere in the permanent gas lines.

The home inspector should keep an eye out for the following conditions:

- **Gas leaks:** Any *strong* gas odor in the house should be dealt with immediately. Call the gas company for assistance right away and have the customer, homeowner, and others present at the inspection leave the house. Don't take any chances in a serious situation. But don't push the panic button unnecessarily either. A faint odor of gas at pilot lights is common and isn't a gas leak.

  However, the faint odor of gas along piping runs or at connections to appliances such as the stove or water heater is an indication of a gas leak. This should be reported as a **safety hazard** with the recommendation that the proper personnel be called in to correct the situation.

- **Improper materials:** Note any gas piping materials used in violation of your own local codes.

- **Poor piping support:** It's important that gas piping be properly supported and not experience any strain against the piping.

- **Missing shut-off valves:** There should be shut-off valves for all gas-fired appliances, and they should be located in the same room as the appliance.

INSPECTING GAS LINES

- Gas leaks
- Improper piping materials or support
- Missing shut-off valves
- Improper tank location

Update:

Corrugated stainless steel tubing (CSST) is allowed in interior and underground locations in accordance with manufacturer's instruction.

For Your Information

Find out what your local codes require for bottled or natural gas piping so you can advise customers of violations.

GAS ODOR

If the home inspector detects a strong odor of gas in a home, he or she should have people leave the house immediately. Call the gas company for help.

- **Improper tank location:** Communities have requirements for the location of LP gas tanks. Because of its explosive nature, LP gas tanks must never be located in the house, basement, or garage. They shouldn't be near an outside window, and, in general, are required to be a minimum of 10' from the house depending on the size of the tank. Violations should be reported as a **safety hazard**.

EXAM

A Practical Guide to Inspecting Plumbing has introduced you to a lot of detail about plumbing. Take the time to test yourself and see how well you've learned it.

To receive Continuing Education Units:
Complete the following exam by filling in the answer sheet found at the end of the exam. Return the answer sheet along with a $50.00 check or credit card information to:

American Home Inspectors Training Institute
N19 W24075 Riverwood Dr., Suite 200
Waukesha, WI 53188

Please indicate on the answer sheet which organization you are seeking CEUs.

It will be necessary to pass the exam with at least a 75% passing grade in order to receive CEUs.

Roy Newcomer

Name_____

Address_____

_____

Phone:_____

e-mail:_____

Credit Card #:_____

Exp Date:_____

Fill in the corresponding box on the answer sheet for each of the following questions.

1. Which action is required by most standards of practice during the plumbing inspection?

   A. Required to state the effectiveness of anti-siphon devices
   B. Required to observe onsite water supply quantity and quality
   C. Required to describe water supply and distribution piping materials
   D. Required to operate automatic safety controls

2. Which of the following does <u>not</u> have to be inspected during the plumbing inspection?

   A. Water heating equipment
   B. Fire and lawn sprinkler systems
   C. Fixtures and faucets
   D. Supply and DWV piping

3. What is the overall purpose of the plumbing inspection?

   A. To determine functional flow and drainage
   B. To identify piping materials
   C. To operate all plumbing fixtures
   D. To identify major deficiencies in the plumbing system

4. What type of piping is <u>most commonly</u> used today as the main service pipe from public water supplies?

A. Copper
B. Galvanized steel
C. Lead
D. PE or PB

5. What is the depth that marks the difference between a shallow well and a deep well?

A. 15'
B. 25'
C. 35'
D. 45'

6. The presence of a pump and a pressure tank in the basement <u>always</u> indicates a private water supply.

A. True
B. False

7. For most communities, how far should a private well be located from a septic system?

A. 18'
B. 30'
C. 50'
D. 75'

8. Identify the following components of a private well system, marked 1, 2, 3, 4:

A. Main shut off, pressure switch, pressure tank, gauge

B. Gauge, tank, switch, shutoff

C. Switch, tank, gauge, shutoff

D. Switch, tank, shutoff, gauge

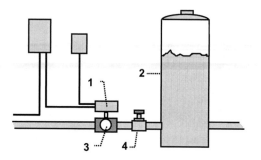

9. What condition causes a well pump to run constantly?

A. A blown fuse
B. A defective motor bearing
C. A waterlogged pressure tank
D. Undersized incoming pipe

10. When should the home inspector recommend a well technician be called to examine the well equipment?

A. If the pressure tank is leaking
B. If the main shut-off valve is missing
C. If the pump is overly noisy
D. If the power to the pump is off

11. What type of supply piping has wiped joints?

A. Copper
B. Yellow Brass
C. Galvanized steel
D. Lead

12. Which type of plastic piping is blue-gray?

A. PB
B. ABS
C. CPVC
D. PVC

13. Which type of supply piping is susceptible to rusting from the inside?

A. Copper
B. Red brass
C. Galvanized steel
D. Lead

14. What type of corrosion is shown in Photo #6?

A. Chemical
B. Galvanic

15. What caused the corrosion in Photo #6?

A. An improper connection between 2 dissimilar metals in the run
B. Improper soldering techniques

16. What is water hammer?

    A. A tool that plumbers use
    B. Greenish stains on copper pipes
    C. Shock waves that occur in the pipes when water is turned off suddenly
    D. Loose pipe supports that rub against the pipes

17. Which of the following is a frost-free hose bib?

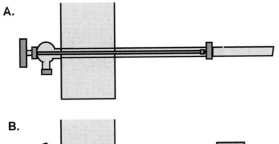

A.

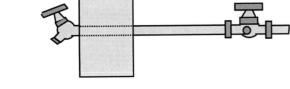

B.

18. How far apart should uninsulated hot and cold water pipes be installed from each other?

    A. 3"
    B. 4"
    C. 5"
    D. 6"

19. What abbreviation is used when measuring water flow rate?

    A. Psi
    B. Gpm
    C. Sq. ft.
    D. DWV

20. Should the home inspector test the valve shown in Photo #7?

    A. Yes
    B. No

21. What is a cross connection as it relates to plumbing?

    A. A dielectric connector
    B. When dissimilar metals are used in a single plumbing run
    C. When waste water can be siphoned into the water supply system
    D. Low water pressure in the supply system

22. What is a soil pipe?

    A. A pipe that carries sewer gas
    B. A pipe that carries waste from toilets
    C. A pipe that carries waste from sinks
    D. A pipe that carries water to the garden

23. What item shown in this photo may not be allowed in some communities?

    A. S-traps
    B. P-traps
    C. Air admittance valve
    D. Double traps at sink

24. What is one of the functions of venting in the plumbing system?

    A. To allow sewer gases to escape
    B. To provide fresh air in bathrooms
    C. To prevent air from entering the plumbing system
    D. To carry waste to the septic tank

25. What size DWV piping is required for the soil stack?

    A. 1" or 2"
    B. 2" or 3"
    C. 3" or 4"
    D. 4" or 5"

26. With horizontal cast iron DWV piping:

    A. The top of the pipe can rust from water sitting in the pipe.
    B. The top of the pipe can split.
    C. Joints can break from the weight.
    D. Sewer gas can eat through the top.

27. Identify each type of trap shown here, 1, 2, 3:
    A. Drum trap, S-trap, P-trap
    B. Drum trap, P-trap, S-trap
    C. P-trap, S-trap, drum trap

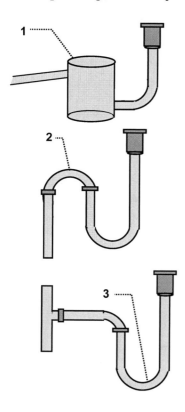

28. What condition in this photo should be reported as a safety hazard?
    A. Possible cross connection
    B. Plastic and metal piping in the same run
    C. The use of a P-trap
    D. The use of rubber hoses for plumbing

29. What findings from the inspection of DWV piping should be reported as a safety hazard?
    A. Hole in the vent pipe, cross connection, improperly installed venting pipes
    B. Use of plastic piping, hole in venting pipes, cross connection
    C. Hole in venting pipe, cross connection, use of copper piping
    D. Hole in vent pipe, improperly installed venting pipes, use of galvanized piping

30. Which type of pump is used to discharge waste from a basement toilet that is below the level of the sewer line?
    A. A sump pump
    B. A well pump
    C. A solid waste pump

31. Which type of water heater has the slowest recovery rate?
    A. Gas
    B. Oil
    C. Electric

32. What is the purpose of the relief valve on a water heater?
    A. To allow water and sediment to be drained from the tank periodically
    B. To allow water and steam to escape when pressure or temperature gets too high
    C. To allow cold water to enter to the bottom of the tank
    D. To allow the water supply to be turned off

33. Why should the extension to the relief valve on a water heater not be threaded at the bottom?
    A. To prevent anyone from getting sprayed in the face when the relief valve trips
    B. To prevent the extension from draining into a laundry tub
    C. To prevent steam from being released from the extension
    D. To prevent the end from being capped.

34. What is NOT the proper installation of a single walled gas water heater flue pipe?
    A. It should run horizontally between the water heater and the chimney
    B. If single-walled, the flue pipe should be kept at least 6 in. from combustibles
    C. It should bent at least 2 ft. above the roof and at least 10 ft. away from any structures on the roof.

35. What happens if inlet and outlet pipes are reversed on the water heater?

    A. Cold water is sent to the bottom of the tank.
    B. Cold water is sent to the top of the tank.
    C. Hot water exits from the top of the tank.
    D. Hot water exits from the anode rod.

36. A thumping or rumbling noise heard while a gas water heater is firing indicates:

    A. A gas leak
    B. A faulty thermostat
    C. Sediment in the tank
    D. A corroded heating chamber

37. What is flame rollout in a gas water heater?

    A. When the flame burns blue and orange
    B. When the flame burns yellow
    C. When the flame burns out from behind the burner plate to the outside of the tank
    D. When the pilot light is out

38. Identify the components of the water heater 1, 2, 3, 4:

    A. Dip tube, relief valve, anode rod, heating element
    B. Anode rod, relief valve, dip tube, heating element
    C. Dip tube, anode rod, relief valve, heating element

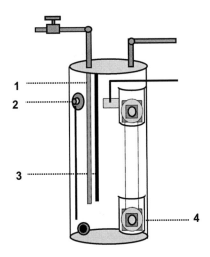

39. How can you test if both heating elements on the electric water heater are working?

    A. Watch the flame in the heating chamber.
    B. Press your thumb against the tank at each element.
    C. Listen to the tank.
    D. Test the faucets for hot water.

40. How can you test if there is a downdrafting condition with a gas water heater?
    A. Watch the flame in the heating chamber.
    B. Trip the relief valve.
    C. Watch the lean of a match near the draft hood.
    D. Smell around the gas control valve and at unions and connections in the gas lines.

41. **Case study:** During a home inspection, you find the water heater as shown in Photo #23. What, if any, finding should be made on the relief valve?
    A. Relief valve is missing and cross connection at laundry tub
    B. Extension has a right angle to it an bottom of extension is threaded and has an elbow
    C. Bottom of extension is threaded and has an elbow, cross connection at laundry tub, and copper piping
    D. Cross connection at laundry tub and extension has right angle to it, PVC piping

42. For the case study shown, what should be reported as safety hazard?
    A. Relief valve is missing, cross connection at laundry tub
    B. Extension has a right angle to it and bottom of extension is threaded
    C. Bottom of extension is threaded and has an elbow and a cross connection at laundry tub
    D. Cross connection at laundry tub and extension has right angle to it

43. A home inspector is required to:
    A. Operate all valves in the plumbing system.
    B. Operate all faucets in the plumbing system except winterized hose bibs.
    C. Trip the relief valve on the water heater.
    D. Turn off the main shut-off valve before beginning the plumbing inspection.

44. If a faucet drips when turned off, it's most likely caused by:

   A. The faucet not secured to the fixture.
   B. A faulty valve or cartridge.
   C. Water hammer.
   D. A deteriorated washer or O-ring.

45. Functional flow at a faucet is defined by:

   A. Whether water drains completely and fast enough.
   B. Whether enough water comes out of the faucet fast enough.
   C. Whether water is hot enough or too hot.
   D. Whether the faucet has an aerator.

46. What is a sign that an old lead or tin shower pan may be present?

   A. A ceramic tiled shower floor
   B. Ceramic tiled shower walls
   C. The presence of a drum trap
   D. Wood rot at the shower door.

47. Why should the home inspector lift or nudge the toilet bowl from side to side?

   A. To check the date of manufacture
   B. To check if there is wood rot around the toilet base
   C. To check if the toilet bowl is cracked
   D. To check if the toilet bowl is tightly secured to the floor

48. What can be the cause of the toilet tank constantly sending water into the bowl?

   A. Vent stoppage
   B. Too little water volume
   C. A defective flush mechanism
   D. Improper slope of a waste pipe

49. What material is commonly used for gas piping in the house?

   A. Copper
   B. Iron or steel
   C. Galvanized steel

50. What should the home inspector do when a faint odor of gas is detected at a standing pilot type burner on the kitchen stove?

   A. Have everyone leave the house and call the gas company for help.
   B. Record the gas odor as a safety hazard.
   C. Recommend that the gas company be called to investigate the odor.
   D. Explain to a customer that a faint gas smell at a pilot light is common and isn't a gas leak.

GLOSSARY

ABS An abbreviation for acrylonitrile butadiene styrene, a black plastic piping used in DWV piping. Uses mechanical no-hub joints.

Aerobic bacteria Bacteria that live in an oxygen-rich environment and break down solid waste matter.

Anaerobic bacteria Bacteria that live in an oxygen-free environment and break down solid waste matter.

Anode rod In a water heater, a rod that will give up ions before the steel tank does, thus protecting the tank. Usually magnesium or aluminum.

Anti-siphon device See *Backflow preventer*.

Aquifer A water-bearing strata in the earth of permeable rock, sand, or gravel.

Artesian well A well whose aquifer has enough pressure to bring water to the surface without a pump.

Backflow preventer A device such as an anti-siphon device or vacuum breaker that prevents water from being siphoned from the supply system into the DWV system.

Branch lines Water supply pipes that carry water from risers to fixtures and faucets in the house.

Brass An alloy of copper and zinc.

Brazed The use of silver solder in the soldering process.

Centrifugal pump A motorized pump that lifts water from a well by means of a paddle wheel driven by a motor. Used in jet and submersible pumps.

Cesspool A masonry lined hole used to hold and break down solid materials from the home's waste system before releasing them into the ground through porous masonry.

Check valve On a sanitary pump, a valve that prevents waste from the sewer line from backing into the crock.

Chemical corrosion Corrosion that occurs when metals react with oxygen, carbon dioxide, or salts in water, using metal atoms to form new compounds.

Cleanout A plug-sealed extension at the sewer outlet which is unplugged when work needs to be performed on the sewer lines.

CPVC An abbreviation for chlorinated polyvinyl chloride, a white or beige plastic piping used in supply piping. Joints are solvent welded.

Cross connection A condition where water from the DWV system can be siphoned back into the supply system.

Desulfovibrio bacteria Bacteria that live in hot water in the presence of magnesium, causing hot water to smell like rotten eggs. Can contaminate water heaters.

Dip tube In a water heater, the tube that sends cold water to the bottom of the tank.

Draft hood A device on top of a water heater that prevents backdrafts from the chimney sending exhaust gases into the home.

Drain pipes Any discharge pipes in the DWV system.

Drain valve On a water heater, a valve at the base of the tank used to drain water.

Drain, waste, and vent system Piping that carries water and waste from home's fixtures to the sewer or septic system. Abbreviated *DWV*.

Drum trap A canister-shaped unvented trap used under fixtures such as bathtubs and laundry sinks.

Dry vent A vent in the home's DWV system that carries only air and water vapor.

Drywell A buried gravel pit that accumulates water and allows it to seep into the ground slowly.

DWV See *Drain, waste, and vent system*.

Electrolytic corrosion Corrosion that occurs when two dissimilar metals are connected to each other in water containing dissolved salts, releasing metal ions and causing a current to flow.

Faucet A device to turn water on and off at a delivery point, which operates with a washer, cartridge, ball, valve, or an O-ring mechanism.

Fixture Any sink, tub, shower, and toilet to which water is delivered in the home.

Flame rollout A condition in a gas water heater where the flames burn outside the heating chamber.

Flue pipe The exhaust pipe running from a gas water heater to the chimney.

Flux A material, which eliminates chemical contamination, used in the soldering process.

Functional drainage A determination of whether water drains fast enough and completely.

Functional flow A determination of whether water flows with enough pressure and volume.

Galvanic corrosion See *Electrolytic corrosion*.

Galvanized steel Steel coated with zinc.

Gpm An abbreviation for gallons per minute, a measurement of water flow.

Gray water Drainage from the clothes washer, laundry tub, or floor drain.

Heating chamber In a gas water heater, the chamber where the gas is burned.

Heating elements In an electric water heater, the upper and lower electrical heating units.

Hose bib A faucet on the exterior of the home.

House main See *Service pipe*.

House trap A large U-shaped fitting with capped heads found where the home waste lines join the public sewer.

ID An abbreviation for inside diameter, referring to pipe measurements.

Jet pump A motorized pump that recirculates pressurized water into a well and pushes water to the surface by means of a centrifugal pump and jet assembly.

Leach field See *Seepage field*.

LP gas Liquified propane or petroleum gas used as a fuel source.

Main shut-off valve A valve that turns off the water supply to the house.

Mechanical vent A device installed for venting purposes in an unvented location. Does not vent to the outside.

Metal shower pan A shower pan made of lead or tin notorious for leaking,

Operating controls On a gas water heater, a control unit with a thermostat and gas controls.

Overflow An auxiliary drain in bathtubs and some sinks.

PB An abbreviation for polybutylene, a blue plastic piping used in supply piping. Uses press-on fittings on joints.

PE An abbreviation for polyethylene, a plastic piping sometimes used in public water systems.

Piston pump A motorized pump that lifts water from a well with a series of pistons. Also called a reciprocating pump.

Pressure gauge On a pressure tank, a gauge that shows the pressure reading of the tank.

Pressure reducing valve A valve on the service pipe near the meter that reduces city water pressure as it reaches the house.

Pressure switch On a pressure tank, a switch that automatically starts and stops the well pump at preset pressures.

Pressure tank A storage tank that holds well water under pressure by being partially filled with air.

Psi An abbreviation for pounds per square inch, a measurement of water pressure.

P-trap A P-shaped trap commonly used today below fixtures and floor drains. Usually vented.

Pump house A separate building built for the purpose of housing well equipment.

PVC An abbreviation for polyvinyl chloride, a white plastic piping used in DWV piping. Uses mechanical no-hub joints.

Reciprocating pump See *Piston pump*.

Relief valve On a water heater, a valve that releases water when temperature or pressure is too high.

Relief valve extension Discharge piping from the relief valve on a water heater to the floor.

Reverse trap toilet A 2-piece (tank and bowl) toilet with a larger wetted area than earlier models.

Risers Water supply pipes that carry water vertically up through the house.

Sanitary pump A sealed crock located in the basement floor containing an electric pump which pumps gray water from fixtures into the sewer line or drywell.

Seepage field Underground porous concrete drain tiles receiving liquid discharge from a septic system for seepage into the ground.

Septic tank A watertight underground tank of concrete, steel, or fiberglass into which household waste is held and broken down for release to a seepage field.

Service pipe The pipe that brings water from its public or private source into the house.

Siphon jet toilet A reverse trap toilet with a quiet flush.

Siphon vortex toilet A late model 1-piece toilet with a large wetted area and a silent flush.

Soil pipes Waste pipes that carry waste from toilets in the home.

Soil stack Main vertical waste pipe fed by all other waste pipes that carries waste to the sewer to septic system.

Soldering The process of using flux and a soft solder to make a joint in copper piping.

Solid waste pump A sealed tank in the basement floor containing an electric pump which pumps toilet waste up to the sewer line.

Stacks Vertical piping in the DWV system.

Standpipe An outdoor elevated water reservoir.

S-trap An S-shaped unvented trap once used under plumbing fixtures.

Submersible pump A motorized pump that sits in the well and pushes water to the surface by means of a motor and centrifugal pump.

Sump pump A pit located in the basement floor containing an electric pump which pumps water from the perimeter drain system away from the house.

Supply system Distribution piping from the source of water supply to the home's fixtures and faucets.

Tankless coil A coil inserted into a boiler to heat water for domestic use.

Trap A device that holds water in the plumbing system and prevents the backflow of gases.

Turn-off valves Valves in the water supply system that allow water to be turned off to certain locations in the house.

Vacuum breaker See *Back-flow preventer*.

Vault An elevated indoor water reservoir, often located in the attic, from which water flows by gravity.

Vent pipes Pipes that carry gases and pressure that builds up in the DWV system.

Vent stack Main vertical vent pipe fed by all other vent pipes that exhausts through the roof.

Vent system All vent pipes in the home, exhausting above the roof.

Washdown toilet An older model 2-piece (tank and bowl) toilet with a large bulge in front of the bowl and a small wetted area.

Waste pipes Drain pipes that carry water away from fixtures in the home.

Water hammer A condition that occurs when water is shut off suddenly and shock waves move back and forth in the supply piping. Can cause banging in loose pipes.

Water heater A steel tank lined with glass, porcelain, or cement in which water is heated for domestic use.

Waterlogged A condition with a pressure tank where its air is absorbed into the water it holds causing a well pump to cycle rapidly and repeatedly.

Well pit An underground chamber, usually made of masonry, that houses well equipment.

Well pump A pump that draws water from a well and pushes it through the home's piping system.

Wet vent A vent pipe in the home's DWV system that combines carrying air and water vapor with carrying waste matter.

Wiping The process of using molten lead to make a bulb-shaped joint in lead piping.

INDEX

Standards of practice 1
Codes
    Plastic piping 27
    Relief valve extensions 60
    Septic systems 10, 42
    S-traps 44
    Water heater exhaust 61
    Water heater location 61
    Wells 10
Corrosion 23, 24
Cross connections 36, 37
Drainage
    Sanitary pumps 54, 57
    Solid waste pumps 55
    Sump pumps 53, 57
DWV system
    Drainage pumps 52-55
    Functional drainage 52
    Inspection 52-55
    Overview 40
    Piping materials 45-47
    Reporting 55-57
    Septic systems 41, 42
    Venting system 43, 44
Fixtures and faucets
    Faucets types 72
    Inspection 72-74
    Reporting 80
    Sinks, tub, showers 74-77
    Toilets 77-78
Functional drainage 52
Functional flow 34-36
Gas piping 81
Hose bibs 32, 33
Hot water system
    Components 59-61
    Electric 68, 69
    Gas 62-67
    Inspection 59-69
    Reporting 70

Inspection
    Standards 1
    DWV system 40-55
    Fixtures and faucets 72-79
    Flow 79
    Gas piping 81, 82
    Hot water system 59-69
    Overview 2-4
    Report 20, 21
    Reporting 19-21, 35-38, 51-
53, 55-57, 64, 70, 73, 80
        Water supply entrance 15-19
        Water supply piping 37, 38
Piping materials
    ABS 28, 44, 45
    Brass 25, 43-45
    Cast iron 45, 46
    Copper 5, 6, 26, 27, 46,
    CPVC 28
    Galvanized steel 6, 26
    Lead 6-8, 25, 26, 46-47
    PB 6, 27
    PE 6
    PVC 28, 47
Private water supply 10-15
Public water supply 5-9
Sanitary pumps 54, 55
Septic systems 41, 42
Showers 74-77
Sinks 74-75
Solid waste pumps 55
Sump pumps 53
Tankless coil 69
Toilets 77, 78
Tubs 74, 75
Valves and devices
    Anti-siphon device 36, 37
    Check valve 54
    Drain valve 61
    Drum trap 45

Gas turn-off 62
Main shut-off 6, 7
Mechanical vent 44
Pressure gauge 14, 15
Pressure reducing valve 6
Pressure switch 14, 15
P-trap 43, 44
Relief valve 60
S-trap 44, 45
Turn-off valve 60
Vacuum breaker 36, 37
Venting system 43-45
Waste disposal 41, 42
Water heaters
    Electric 68, 69
    Gas 62-67
Water supply entrance
    Inspection 7-9, 15-19
    Piping materials 5, 6
    Private supply 10-15
    Public supply 5-9
    Pump components 14
    Types of well pumps 10-13
    Reporting 19-21
Water supply piping
    Corrosion 23, 24
    Cross connections 36, 37
    Functional flow 34-36
    Hose bibs 32, 33
    Inspection 29-34
    Overview 23
    Piping materials 25-28
    Reporting 37, 38
Well pumps
    Inspecting 15-19
    Jet 11, 12
    Piston 11
    Submersible 12, 13

A Practical Guide to Inspecting Program
Study Unit Four, Inspecting Plumbing

Student Name: _____ Date: _____

Address: _____

Phone: _____ Email: _____

Organization obtaining CEUs for: _____ Credit Card Info: _____

After you have completed the exam, mail *this exam answer page* to American Home Inspectors Training Institute. You may also fax in your answer sheet. You will be notified of your exam results.

Fill in the box(es) for the correct answer for each of the following questions:

1. A☐ B☐ C☐ D☐	24. A☐ B☐ C☐ D☐	47. A☐ B☐ C☐ D☐
2. A☐ B☐ C☐ D☐	25. A☐ B☐ C☐ D☐	48. A☐ B☐ C☐ D☐
3. A☐ B☐ C☐ D☐	26. A☐ B☐ C☐ D☐	49. A☐ B☐ C☐
4. A☐ B☐ C☐ D☐	27. A☐ B☐ C☐	50. A☐ B☐ C☐ D☐
5. A☐ B☐ C☐ D☐	28. A☐ B☐ C☐ D☐	
6. A☐ B☐	29. A☐ B☐ C☐ D☐	
7. A☐ B☐ C☐ D☐	30. A☐ B☐ C☐	
8. A☐ B☐ C☐ D☐	31. A☐ B☐ C☐	
9. A☐ B☐ C☐ D☐	32. A☐ B☐ C☐ D☐	
10. A☐ B☐ C☐ D☐	33. A☐ B☐ C☐ D☐	
11. A☐ B☐ C☐ D☐	34. A☐ B☐ C☐	
12. A☐ B☐ C☐ D☐	35. A☐ B☐ C☐ D☐	
13. A☐ B☐ C☐ D☐	36. A☐ B☐ C☐ D☐	
14. A☐ B☐	37. A☐ B☐ C☐ D☐	
15. A☐ B☐	38. A☐ B☐ C☐	
16. A☐ B☐ C☐ D☐	39. A☐ B☐ C☐ D☐	
17. A☐ B☐	40. A☐ B☐ C☐ D☐	
18. A☐ B☐ C☐ D☐	41. A☐ B☐ C☐ D☐	
19. A☐ B☐ C☐ D☐	42. A☐ B☐ C☐ D☐	
20. A☐ B☐	43. A☐ B☐ C☐ D☐	
21. A☐ B☐ C☐ D☐	44. A☐ B☐ C☐ D☐	
22. A☐ B☐ C☐ D☐	45. A☐ B☐ C☐ D☐	
23. A☐ B☐ C☐ D☐	46. A☐ B☐ C☐ D☐	